JN228412

フクシマについて、
お案じの向きには、私から保証をいたします。
状況は、統御されています。
東京には、
いかなる悪影響にしろ、
これまで及ぼしたことはなく、
今後とも及ぼすことはありません。

——第125回国際オリンピック委員会における安倍晋三首相によるプレゼンテーションでの発言（2013年9月7日）

Some may have concerns about Fukushima. Let me assure you,the situation is under control. It has never done and will never do any damage to Tokyo.

—Remarks in the presentation by Prime Minister Shinzo Abe at the 125th International Olympic Committee（September 7,2013）

IOC総会における安倍総理プレゼンテーション-平

27.180 回視聴

年9月7日

高評価

低評価

真実から目を
犯罪である。

It is a crime to take your eyes off the truth.

逸らすことは

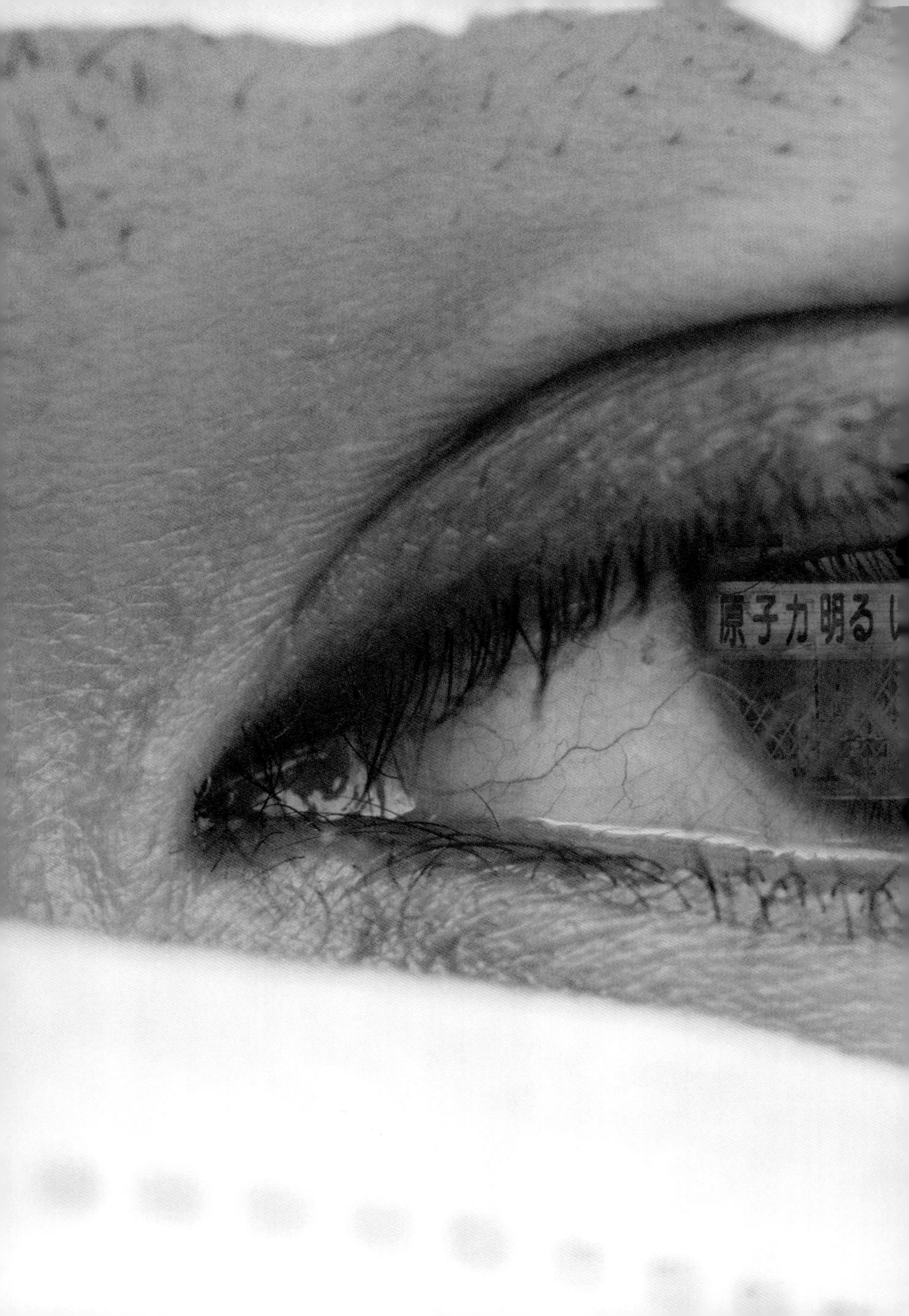
原子力明る

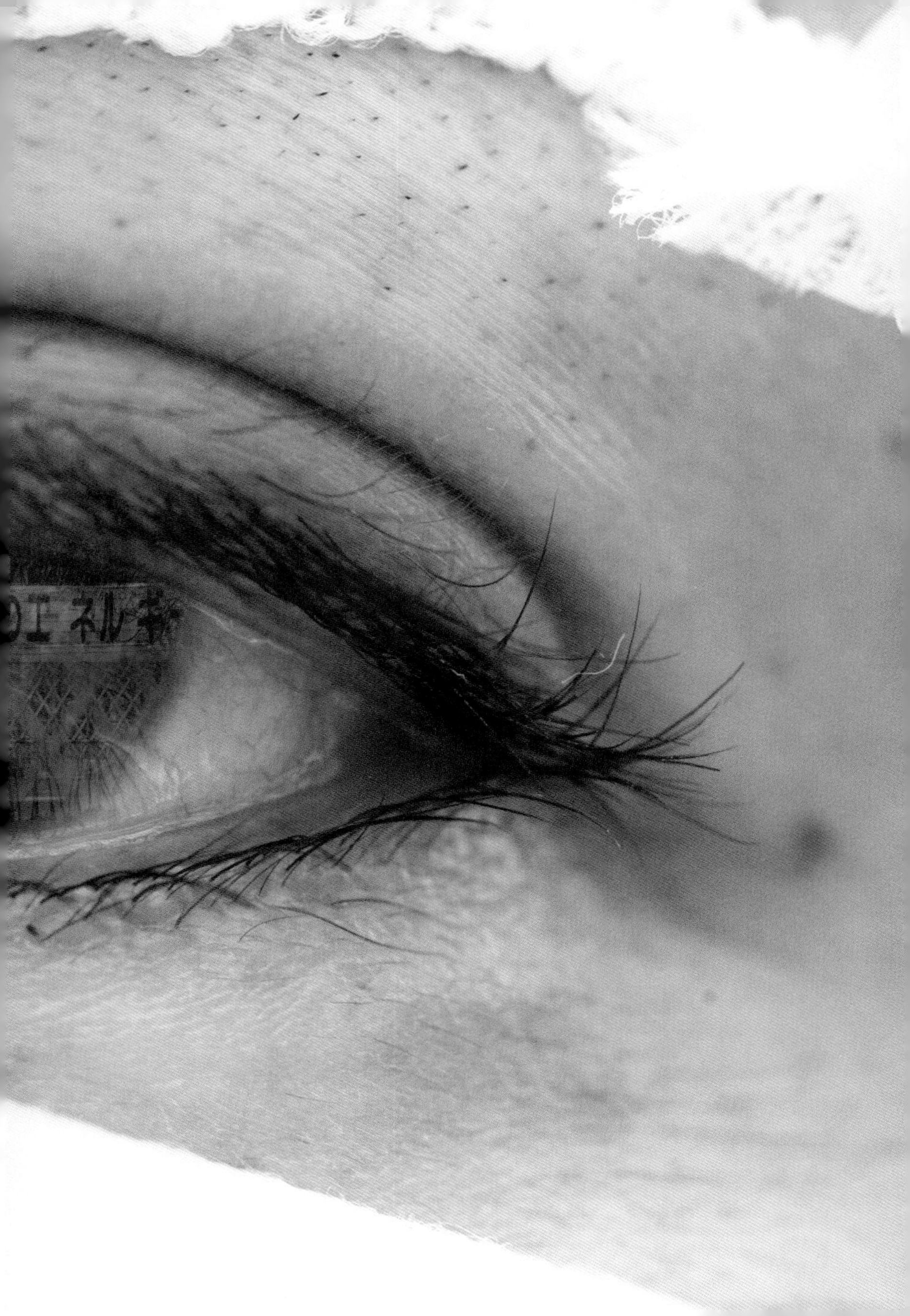

本書について

2018年7月、小出裕章氏はひとりの日本人女性からの依頼を受け、「フクシマ事故と東京オリンピック」と題する文章を書いた。その後それは英訳され、同年10月、世界各国のオリンピック委員会などに書簡として送られた。

本書はその原稿を基に一部加筆・修正したものである。

―― 編集部

About this book

In July 2018, the author Hiroaki Koide took upon a request of a Japanese woman and wrote "The Fukushima Nuclear Disaster and the Tokyo Olympics". After being translated into English, this was sent as a letter to organizations such as International Olympic Committee in October 2018.
This book has been partially revised and corrected based on this original document.
——Editors

[ドイツ語／German]

Zu diesem Buch

Im Juli 2018 kam Koide Hiroaki der Bitte einer Japanerin nach, einen Text über das Problem “Der Unfall in Fukushima und die Olympischen Spiele in Tōkyō“ zu schreiben. Dieser Text wurde daraufhin ins Englische übersetzt und im Oktober 2018 per Brief an alle Nationalen Olympischen Komitees der Welt versandt.

Das vorliegende Buch beruht auf diesem Manuskript, das ergänzt und überarbeitet wurde.

—Die Herausgeber

[フランス語／French]

A propos de ce livre

En juillet 2018, Hiroaki Koide a écrit un texte intitulé "L'accident de Fukushima et les Jeux Olympiques de Tokyo" à la demande d'une femme japonaise. Après cela, il a été traduit en anglais et envoyé sous forme d’une correspondance aux Comités des Jeux Olympiques de pays membres en octobre de la même année.

Ce livre a été partiellement révisé et corrigé sur la base de cette correspondance.

—Le Département éditorial

[スペイン語／Spanish]

Acerca de este libro

En julio de 2018, Koide Hiroaki escribió un texto titulado “El accidente de Fukushima y las Olimpíadas de Tokio”, tras recibir un encargo de parte de una conocida japonesa. Luego, en octubre del mismo año, ese texto fue traducido al inglés y enviado a los comités olímpicos de diferentes países del mundo.

Este libro es una revisión y una reescritura de dicho texto.

—Los editors

[ロシア語／Russian]

О книге

В июле 2018 года Коидэ Хироаки по просьбе одной японской женщины написал статью «Авария на АЭС Фукусима-1 и Олимпийские игры в Токио». Затем она была переведена на английский язык и в октябре того же года разослана в Национальные олимпийские комитеты стран мира.

Данная книга – исправленная и дополненная версия этой статьи.

—Редакция

[中国語／Chinese]

关于本书

2018年7月，小出裕章先生受一位日本女性委托，撰写了《福岛事故与东京奥运会》一文。随后，该篇文章被翻译成英文，并于同年10月作为书信发送至世界各国的奥委会等组织。

本书在该原文的基础上，进行了部分补充及修正。

——编辑部

[アラビア語／Arabic]

حول هذا الكتاب

في الشهر السابع من عام 2018 قام السيد كويده هيروأكي بكتابة مقال بعنوان "حادث فوكوشيما وأولمبياد طوكيو" وذلك بناء على طلب من إحدى النساء اليابانيات. وبعد ذلك تمت ترجمة المقال إلى الإنكليزية، وفي نفس العام من الشهر العاشر تم إرساله كرسالة إلى لجنة الأولمبياد في كل دول العالم.

وهذا الكتاب مبني على هذا المقال مع بعض الإضافات والتعديلات.

— قسم التحرير

目次

Contents

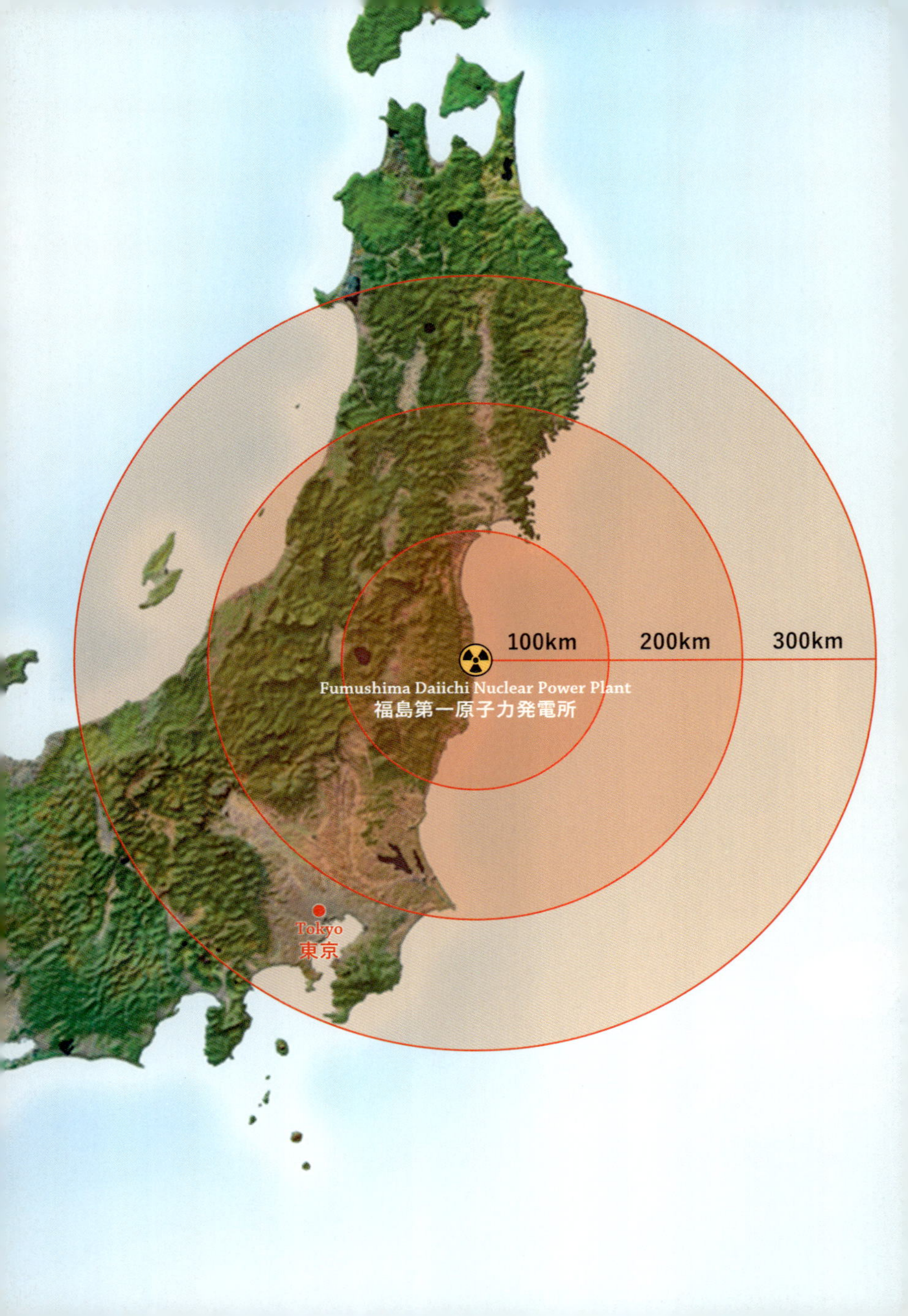
100km
200km
300km
Fumushima Daiichi Nuclear Power Plant
福島第一原子力発電所
Tokyo
東京

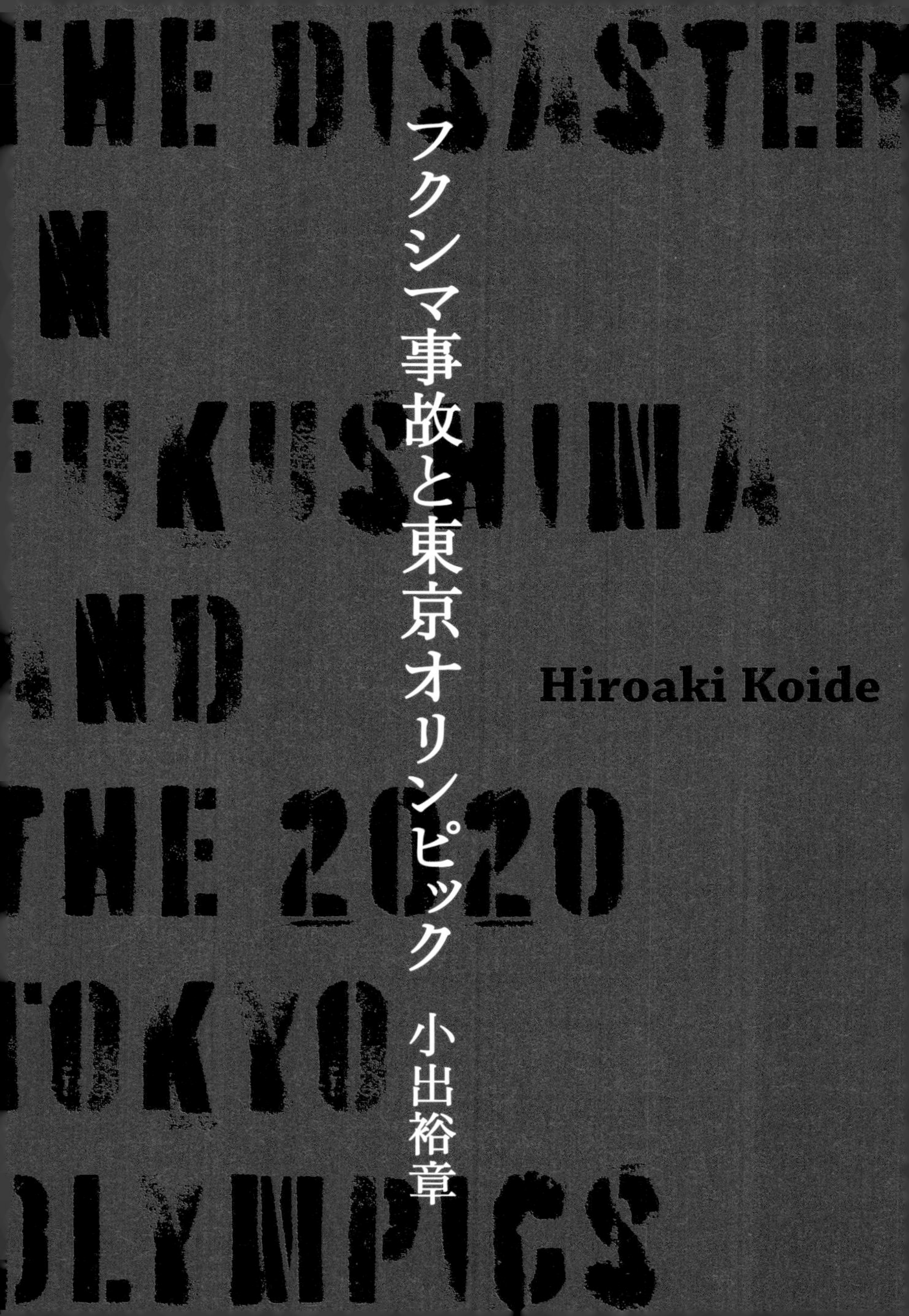
THE DISASTER
IN
FUKUSHIMA
AND
THE 2020
TOKYO
OLYMPICS
フクシマ事故と東京オリンピック
Hiroaki Koide
小出裕章

2011年

On March

3月11日、

1th 2011,

2011 MAR

平成23年 昭和86年

MON TUE WED

1

1/27 七赤 先負 う
全国火災予防運動

2

1/28 八白 仏滅 た

あと15

7

8

11

10

14

15

FRI
SAT
3
4
5
29 九紫
ひな祭　耳の日
先週
13
12
10
11
12
16
17
18
19
とと い
きのう
昨日
きょう
今日
今週

重油タンク
ール水サージタンク

東京電力・福島第一原子力発電所は巨大な地震と津波に襲われ、全所停電となった。

全所停電は、「原発が破局的事故を引き起こす一番可能性の高い原因」と専門家は一致して考えていた。

その予測通り、福島第一原子力発電所の原子炉は熔け落ちて、大量の放射性物質を周辺環境にばらまいた。

a severe earthquake struck the Tohoku region in Japan, causing a Tsunami which hit the Pacific coast of Fukushima and Miyagi and Iwate prefectures, causing a power failure at the Tokyo Electric Power Company Holding's (TEPCO), at the Fukushima Daiichi Nuclear Power Plant.

As all scientists know, a total power outage carries the highest risk of a potentially catastrophic incident.

Bearing out their prediction, the cooling system failed and the nuclear reactor cores melted down; thus high levels of radiation were spread to the surrounding environment.

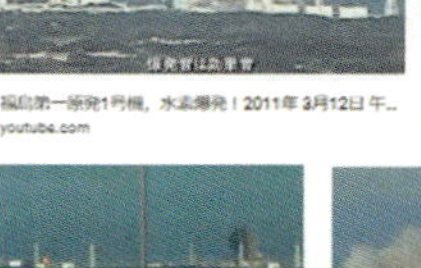

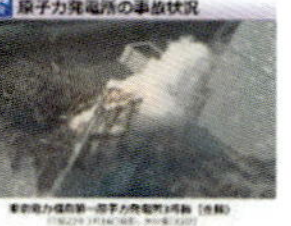

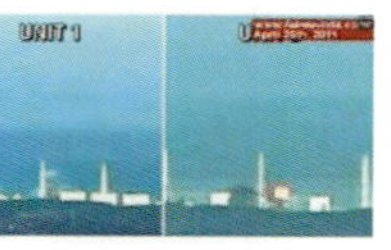

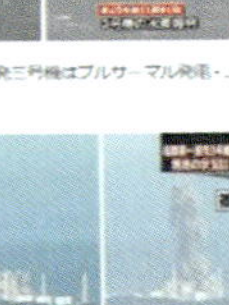

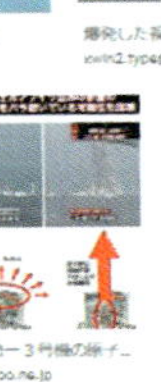

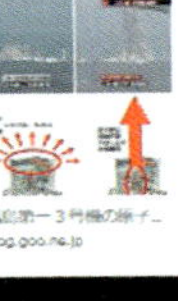

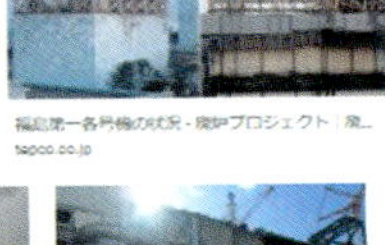

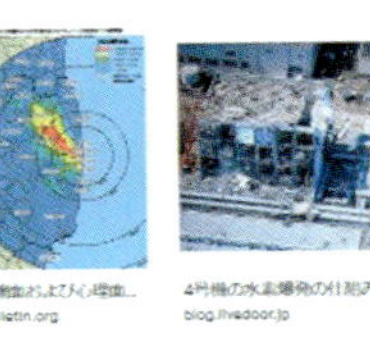

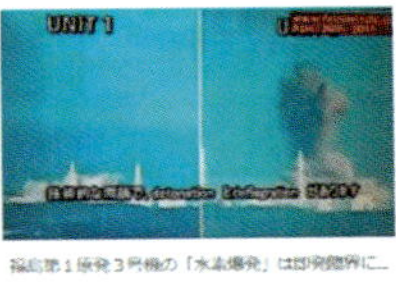

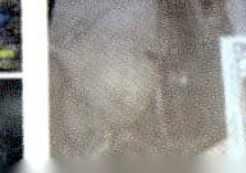

日本国政府が国際原子力機関に提出した報告

広島原爆

のセシウ

気中に放

According to Japanese government's report[*1] delivered to the International Atomic Energy Agency (IAEA), $1.5x10^{16}$ Becquerel (equal to

よると、その事故では、1.5×10の16乗ベクレル、

168発分137が大出された。

168 times the Ceasium-137 of the Hiroshima nuclear bomb) was released as a result of that accident.

HIROSHIMA (ATOMIC) STRIK

広島原爆1発分の放射能でさえも猛烈に恐ろしいものだが、なんとその168倍もの放射能が大気中にばらまかれたと日本政府が言っているのである。

この事故で1、2、3号機の原子炉が熔け落ちたのだが、その炉心の中には、合計で7×10の17乗ベクレル、広島原爆に換算すれば約8000発分のセシウム137が存在していた。そのうち大気中に放出されたものが168発分であり、海に放出されたものを合わせると、現在までに環境に放出されたもの(セシウム137)は、広島原爆約1000発分程度であろう。

The nuclear bomb which was dropped on Hiroshima was just one bomb, and yet had an extremely destructive force.
The cores of Unit 1, Unit 2 and Unit 3 contained 7×10^{17} Becquerel, equal to 8000 times the quantity of Caesium-137 spread by the Hiroshima explosion.
Therefore, it can be estimated that a radioactivity equal to 168 times the power of the Hiroshima bomb was released in the air in the Fukushima event. Some of it wound up in the ocean, making for a total estimate of the radiation of almost 1000 Hiroshima bombs released in the environment from March 11th 2011 until now.

セシウム137はウランが核分裂して生成される核分裂生成物の一種であり、フクシマ事故で人間に最大の脅威を与える放射性物質である。つまり、炉心にあった放射性物質の多くの部分が、いまだに福島第一原子力発電所の壊れた原子炉建屋などに存在している。これ以上、炉心を熔かせば、セシウム137を含む放射性物質が再度環境に放出されてしまうことになる。

Moreover, it is safe to believe that a large part of the radioactivity still remains in those damaged reactors at the Fukushima nuclear power plant. The more we let the cores melt, the more nuclear components, including Caesium-137, will be released into the environment. Caesium-137 is one of the products of nuclear fission, and amongst the many released by the Fukushima accident, it is the most dangerous to human health.

それを防ごうとして、事故から8年以上経った今でも、どこかにあるであろう熔け落ちた炉心に向けてひたすら水を注入している。そのため、毎日数百トンの放射能汚染水が貯まり続けている。

東京電力は敷地内に1000基近いタンクを作って汚染水を貯めてきたが、その総量はすでに100万トンを超えた。[*2]

敷地には限りがあり、タンクの増設にも限度がある。近い将来、東京電力は放射能汚染水を海に流さざるを得なくなる。

TEPCO, which owns Fukushima Daiichi Nuclear Power Plant, does not know at what depth the now-melted cores are exactly and they still keep pouring huge amounts of water to the area in an effort to cool them off, as it is imperative to prevent the cores from melting further and to reduce the further release of radiation.
The constant cooling leads to the contamination of several hundred tons of water per day, and TEPCO has been forced to build approximately 1000 tanks inside the plant to store the contaminated water. The total quantity of contaminated water in those tanks amounted to over one million tons.[*2]
With limited territory at TEPCO's disposal, it is reasonable to believe that in the near future, when tha tanks' clearance is exceeded, the contaminated water will be poured into the ocean.

もちろん一番大切なのは、熔け落ちてしまった炉心を少しでも安全な状態に持っていくことだが、8年以上の歳月が経った今でも、熔け落ちた炉心がどこに、どんな状態であるかすら分かっていない。なぜなら現場に行かれないからである。

事故を起こした発電所が火力発電所であれば、簡単である。当初、何日間か火災が続くかもしれないが、それが収まれば現場に行くことができる。事故の様子を調べ、復旧し、再稼働することだってできる。

しかし、事故を起こしたのが原子力発電所の場合、事故現場に人間が行けば、死んでしまう。

国と東京電力は代わりにロボットを行かせようとしてきたが、ロボットは被曝に弱い。なぜなら命令が書き込まれているICチップに放射線が当たれば、命令自体が書き変わってしまうからである。そのため、これまでに送り込まれたロボットはほぼすべてが帰還できなかった。

Although the best thing to do would be to move the melted cores to a safe, sealed place, since the accident eight years ago, the administration is neither able to exactly locate the cores nor determine their actual condition. Nobody can get near enough the site to investigate. If it were a thermal power plant, and not a nuclear power plant, the accident would be easier to deal with; it could start a fire that would last a few days, but once it died out, we could easily approach the site, assess its conditions and eventually repair the damage and re-start the plant.
Approaching a damaged nuclear site means exposing any person entrusted with this task to high levels of radioactivity and, most probably, to sudden death.
The government and TEPCO did try to send a robot. However the radioactivity is so strong that it interfered with the robot's IC tip's memory, making it unuseless. Out of all the robots they have sent to the site, none has returned.

2017年1月末に、東京電力は原子炉圧力容器が乗っているコンクリート製の台座(ペデスタル)内部に、いわゆる胃カメラのような遠隔操作カメラを挿入した。圧力容器直下にある鋼鉄製の作業用足場に大きな穴が開き、圧力容器の底を抜いて熔け落ちて来た炉心が、さらに下まで落ちていることが分かった。

しかし、その調査ではもっと重要なことが判明した。人間は全身で8シーベルト被曝すれば、確実に死ぬ。圧力容器直下での放射線量は一時間当たり20シーベルトであり、それすら大変な放射線量である。しかし、そこに辿り着く前に530あるいは650シーベルトという放射線が計測された[*3]。そして、この高線量が測定された場所は、円筒形のペデスタルの内部ではなく、ペデスタルの壁と格納容器の壁の間だったのである。

At the end of January 2017, TEPCO managed to insert a remote control camera, similar to a endoscope, inside of the pedestal on which the reactor stands, and found out that the core is leaking through a big hole at the very bottom of the reactor.

On that occasion, they were able to measure a 20Sievert radioactivity level per hour under the reactor, which is an enormous amount of radioactivity, when the maximum threshold for human beings is 8 Sievert. Approaching the reactor, the radiation counter peaked at 530-650 Sievert[*3], and this only two years ago. The place with the highest radioactivity levels was not inside the pedestal but outside, between the pedastal's wall and vessel's wall.

東京電力や国は、熔け落ちた炉心はペデスタルの内部に饅頭のように堆積しているというシナリオを書き、「30年から40年後には、熔け落ちた炉心を回収し容器に封入する。それを事故の収束と呼ぶ」としてきた。

しかし実際には、熔けた核燃料はペデスタルの外部に流れ出、飛び散ってしまっているのである。やむなく国と東京電力は「ロードマップ」を書き換え、格納容器の横腹に穴を開けてつかみ出すと言い始めた。しかし、そんな作業をすれば、労働者の被曝量が膨大になってしまい、出来るはずがない。

After the accident, TEPCO and the government speculated that most of the three cores melted through the reactor and stopped inside the pedestal, lying intact at the bottom of the containment vessel. This would allow recovery of the cores in 30-40 years and strage in other vessels, thus allowing for the declaration of the full and complete resolution of the Fukushima disaster.Contrary to their hypothesis, the melted cores are not intact at all and are scattered outside of the pedestal, making it impossible for them to be recovered as envisaged by the authorities.
TEPCO and the government then had to change their "Road Map". Immediately thereafter they announced a new project according to which, by drilling a hole into the vessel's wall, the core could be reached and removed from its lodging; however, this proposal is not viable because of the amount of radiation that the workers would be exposed to.

私は当初から、旧ソ連チェルノブイリ原子力発電所事故の時にやったように、石棺で封じるしかないと言ってきた。

そのチェルノブイリ原発の石棺は30年たってボロボロになり、2016年11月にさらに巨大な第2石棺で覆われた。その第2石棺の寿命は100年という。その後、どのような手段が可能かは分からない。

As for myself, I used to insist that the reactors affected by the disaster be sealed by the use of a "sarcophagus" such as the one used by the former Soviet government following the Chernobyl disaster. However this sarcophagus became obsolete after thirty years, to the point that it was sealed, in November 2016, with a second bigger cover which is claimed to last up to one hundred years. After that idea, I cannot say which method should be used.

今日生きている
してチェルノ
を見ることはで
フクシマ事故の
ている人間のす
終わりはしない
落ちた炉心を容

人間の誰一人と
り事故の収束
きない。ましてや
収束など今生き
べてが死んでも
もし仮に、熔け
器に封入するこ

とができたとし

て放射能が消

その後数十万

その容器を安

なければならな

After all, no one who lived through the Chernobyl disaster will be alive when and if it ever comes to an end.Given the situation in Chernobyl, let us not even talk about Fukushima; the damage it is causing will extend well beyond the lifetime of the people who wit-

も、それによっ
る訳ではない。
から100万年、
に保管し続け
いのである。

nessed it, and even if we managed to move the melted cores into a sealed and safe container, this is still not the ultimate solution, since covers do deteriorate and situation must be monitored for potentially several hundred thousand up to on million years.

発電所周辺の環境でも、極度の悲劇がいまだに進行中である。

事故当日、原子力緊急事態宣言[＊4]が発令され、初め3キロ、次に10キロ、そして20キロと強制避難の指示が拡大されていき、人々は手荷物だけを持って家を離れた。家畜やペットは棄てられた。

そしてさらに、福島第一原子力発電所から40〜50キロも離れ、事故直後は何の警告も指示も受けなかった飯舘村は、事故後1カ月以上たってから極度に汚染されているとして、避難の指示が出され、全村離村となった。

The tragedy is still in progress in the area around the Fukushima Daiichi Nuclear Plant.
On the day of the disaster, the Japanese government declared the state of nuclear emergency situation(*4). All residents were evacuated to starting with those in a range of 3km from the nuclear plant, only to extend soon afterwards to an area between 10km to 20km. Evacuation procedures were so quick that people were permitted to only bring what was strictly necessary, leaving most of their personal belongings, livestock and even their pets behind.
Iitate-mura is one of the villages in Fukushima prefecture, lying roughly 40-50km away from the Fukushima Daiichi Nuclear Plant. Its inhabitants were not informed of any contamination of their village on that very day of the accident but after one month, they were forced to immediately abandon their village due to the extreme contamination.

30
川井魚店

(株)東北エルピー
ガス設備工事一式
☎31-0333
SK-II
2020
カムラ

重点目標
じぶんのかんがえを
わかりやすくつたえ
学級目標
ともにのびる子ども
教育目標
進んで学習する子
仲よく助け合う子
たくましくやりぬく子
またいつか
この教室で
会いましょう!!
平成二十四年
三月二十六日
まさみ先生より
3月
消火器
修・卒業証書授与式々場

人々の幸せと

いったいどの

言うのだろう

What is happiness?

ま、

ようなことを

多くの人にとって、家族、仲間、隣人、恋人たちとの穏やかな日が、明日も、明後日も、その次の日も何気なく続いていくことこそ、幸せというものであろう。

それがある日突然に断ち切られた。

避難した人々は、初めは体育館などの避難所、次に、2人で四畳半の仮設住宅、さらに災害復興住宅や、みなし仮設住宅へ移動させられた。その間に、それまで一緒に暮らしていた家族はバラバラになった。生活を丸ごと破壊され、絶望の底で自ら命を絶つ人も、未だに後を絶たない[＊5]。

It implies a lot of simple things yet not to be taken for granted, such as living together with families, friends, neighbors and lovers for the entire duration of your life. In Fukushima, however happiness came to an abrupt end, as people unexpectedly lost everything they had. At first the evacuated people were moved to temporary shelters, then to small houses with one small room for two persons. In any event they have been isolated from their former community; families were forced to separate, with fathers remaining in Fukushima to work, and many could not cope with the situation and collapsed. Their former life has been destroyed and people, out of deep despair, never stop taking their own lives [*5].

キケン!

おすすめ品
子供長袖トレーナー・Tシャツ 各種
990円
990
ともに半世紀
490
980
980

新入学
準備
100
カラフルショッピングエコバッグ
EDWIN

それだけではない。極度の汚染のために強制避難させられた地域の外側にも、本来であれば「放射線管理区域」にしなければいけない汚染地帯が広大に生じた。

「放射線管理区域」とは、放射線を取り扱って給料を得る大人、放射線業務従事者だけが立ち入りを許される場である。しかも、放射線業務従事者であっても、放射線管理区域に入ったら、水を飲むことも食べ物を食べることも禁じられる。もちろん寝ることも禁じられる。放射線管理区域にはトイレすらなく、排せつもできない。ところが国は、今は緊急事態だとして、従来の法令を反故にし、その汚染地帯に数百万人の人を棄て、そこで生活するように強いた。

棄てられた人々は、赤ん坊も含めそこで水を飲み、食べ物を食べ、寝ている。当然、被曝による危険を背負わせられている。棄てられた人は皆不安であろう。

Nuclear fall out affects another vast area around the evacuated zones. This is the so-called "Radiation Control Area or RCA".
An RCA is a restricted area where entrance is strictly prohibited except to authorized personnel, generally professionals and nuclear engineers. No one is permitted to drink, eat or sleep inside the area.
However, the Japanese government allows millions of residents to live in areas that have the same characteristics as the RCA's without qualifying these areas as such, in contempt of the law formerly in force.The Japanese government justified this deviation from the applicable law on account of emergency condition.
These residents include infants and children who are forced to live in a place as contaminated as an RCA and are exposed to an adverse environment on a daily basis.

被曝を避けるために、仕事を捨て、家族全員で避難した人もいる。子どもだけは被曝から守りたいと、男親は汚染地に残って仕事をし、子どもと母親だけ避難した人もいる。でも、そうすれば、生活が崩壊したり、家庭が崩壊したりする。汚染地に残れば身体が傷つき、避難すれば心が潰れる。

棄てられた人々は、事故から８年以上、毎日毎日苦悩を抱えて生きている。

それなのに国は、2017年３月になって、一度は避難させた、あるいは自主的に避難していた人たちに対して、１年間に20ミリシーベルトを越えないような汚染地であれば帰還するよう指示し、それまでは曲がりなりにも支援してきた住宅補償を打ち切った。そうなれば、汚染地に戻らざるを得ない人も出てくる。

Some families have decided to quit their jobs and leave Fukushima in order to protect their children from exposure; as we have seen above, others have decided to live apart when husband usually stay behind and continues to work in Fukushima and his wife and children moved out. Fukushima is a conservative social environment; people respect their traditions and have been living there for generations. Such a separation deeply affects the family structure as its members lose their normal lives. They are left with only two painful options: either they stick to their homes and traditions in a now contaminated area, exposing themselves to the risk of damaging their health, or they leave and suffer the traumatic loss of the lives they once knew. These people have been living the hardest life ever for the past eight years and probably will for many years to come.

To add insult to injury, since March 2017 the government has started to relocate Fukushima refugees back to their home towns in the vicinity of the plant where radioactivity levels are below 20mSv per year; any contributions for alternate housing have been revoked and, as a consequence, people are forced to make their ways back to their villages notwithstanding the fact that these places are still heavily contaminated.

今、福島では、復興が何より大切だとされている。

そこで生きるしかない状態にされれば、もちろん皆、復興を願うしかない。それに人は、毎日、恐怖を抱えながらでは生きられない。汚染があることを忘れてしまいたいし、幸か不幸か放射能は目に見えない。国や自治体は、積極的に忘れてしまえと仕向けてくる。そのため、汚染や不安を口にすれば、復興の邪魔だと非難されてしまう。

Nowadays, reconstruction represents the top priority for the Fukushima municipality, and people who do not have any other choice but to live in Fukushima will probably follow suit and will settle in the environs of Fukushima and forget that it still is a dangerous place, rather than living their whole lives in fear and anxiousness. They might even manage to forget radioactive contamination which, (un)fortunately or unfortunately, is invisible. And the government and the local municipality help people forget in any possible way. Should the residents talk about contamination or simply voice their concerns, they will stand acused of being a hurdle to reconstruction.

KOMATSU

2014年11月9日 / Nov. 9, 2014

2016年5月9日 / May. 9, 2016

2015年5月13日 / May. 13,2015

2017年4月18日 / Apr. 18, 2017

１年間に20ミリシーベルトという被曝量は、かつて私がそうであった「放射線業務従事者」に対して国が初めて許した被曝の限度である。それを被曝からは何の利益も受けない人々に許すこと自体、許しがたい。ましてや、赤ん坊や子どもは被曝に敏感であり、彼らには日本の原子力の暴走、フクシマ事故になんの責任もない。そんな人たちにまで、放射線業務従事者の基準を当てはめるなど、決してしてはならないことである。

しかし、日本の国は「今は原子力緊急事態宣言下にあるから、仕方がない」と言う。

緊急事態が丸１日、丸１週間、１ヶ月、場合によっては残念ながら１年ぐらい続いてしまったということであれば、まだ理解できないわけではない。しかし実際には、事故後８年以上経っても「原子力緊急事態宣言」は解除されていない。

It is worth mentioning that after the accident the Japanese government reviewed the legal limit of radiation exposure by raising it to 20mSv/year. It must be said that 20mSv/year was once the limit set only for experts or scientists who worked with nuclear radiation as I did, and was never applied to regular civilians, which makes the new limit totally unacceptable especially for infants and children, who are more sensitive to radiation and who have no responsibility either for corrupt politics or for the Fukushima nuclear disaster.

The Japanese government keeps on saying that there is not much they can do because of the nuclear emergency situations.

A declaration of emergency can last a whole day, a week or even a whole month. In the worst cases it can last up to one year, and I could even understand that. But the declaration of emergency related to the Fukushima accident has not been lifted yet, after over eight years.

国は、意図的にフクシマ事故を忘れさせてしまおうとし、マスコミも口をつぐんでいるから、「原子力緊急事態宣言」が今なお解除できず、本来の法令が反故にされたままであることを多くの国民は忘れさせられてしまっている。

環境を汚染している放射性物質の主犯はセシウム137であり、その半減期は30年。100年経っても、ようやく10分の1にしか減らない。

実は、この日本という国は、これから100年たっても、「原子力緊急事態宣言」下にあるのである。

Apparently, the government actively promotes oblivion about the Fukushima disaster, in cooperation with the mass media which fails to report the real risk of contamination and the actual situation in Fukushima today. Most Japanese are not aware that Fukushima still is under nuclear emergency and that the government is disregarding the application of current laws.
The most harmful element for the environment and health is Caesium-137; it takes 30 years to reduce its quantity by half and after 100 years 10% of it still remains, it is enough to state that 100 years from now Japan will still be under nuclear emergency situations.

オリンピックは、いつの時代も国威発揚に利用されてきた。

近年は、箱モノを作っては壊す膨大な浪費社会と、それにより莫大な利益を受ける土建屋を中心とした企業群がオリンピックを食い物にしてきた。

しかし、今もっとも大切なのは、「原子力緊急事態宣言」を一刻も早く解除できるよう、国の総力を挙げて働くことである。フクシマ事故の下で苦しみ続けている人たちの救済こそ、最優先の課題である。少なくとも、罪のない子どもたちを被曝から守らなければならない。

The Olympic Games have always been used as propaganda to promote people's nationalism.
Building many huge structures without any prospect of use after the Olympics is just big business for companies including general contractors who make huge profits in a consumer society.
I think the Japanese government has to make a huge effort and do its best to lift the declaration of nuclear emergency related to the Fukushima accident and to save Fukushima's residents, who still live in tremendous situations — saving children first if they cannnot act simultaneously.

それにもか
この国はオ
大切だとい

However, the Japanese government considers a successful 2020 Tokyo Olympics as its most important objective.

かわらず、
リンピックが
つ。

TOKYO 2020

内部に危機を抱えれば抱えるほど、権力者は危機から目を逸らせようとする。そして、フクシマを忘れさせるため、マスコミは今後ますますオリンピック熱を加速させ、オリンピックに反対する輩は非国民だと言われる時が来るだろう。

先の戦争の時もそうであった。

マスコミは大本営発表のみを流し、ほとんどすべての国民が戦争に協力した。自分を優秀な日本人だと思っていればいる人ほど、戦争に反対する隣人を非国民と断罪して抹殺していった。しかし、罪のない人を棄民したまま「オリンピックが大切だ」という国なら、私は喜んで非国民になろうと思う。

They need to organize big events like the Olympics to distract people from other serious problems, and they involve the mass media in fomenting “Olympic Fever”, making whoever is against the 2020 Tokyo Olympics seem like a bad citizen.
During the Second War, the Japanese government spread just the information convenient to them for their propaganda. Almost every Japanese citizen cooperated with the government, and “good citizens” accused as “bad citizens” their neighbors who opposed the system, and had them arrested.
If my country considers a successful 2020 Tokyo Olympics as it most important goal, rather than saving innocent citizens, I would prefer to be regarded as a “bad citizen”.

フクシマ事故は、巨大な悲劇を抱えたまま今後100年の単位で続く。

膨大な被害者を横目で見ながら、この事故の加害者である東京電力、政府関係者、学者、マスコミ関係者などは、誰一人として責任を取っていない。処罰もされていない。

それを良いことに彼らは、今は止まっている原子力発電所を再稼働させ、海外にも輸出するとまで言っている。

原子力緊急事態宣言下の国で開かれる東京オリンピック。

それに参加する国や人々は、もちろん一方では被曝の危険を負うが、また一方では、この国の犯罪に加担する役割を果たすことになる。

The truth is that the Fukushima disaster will last over 100 years and, to my utmost surprise, no one has been formally incriminated yet: no TEPCO representatives, no government official, no politician, no specialists, no media representatives who have caused it. Nobody has even taken any responsibility for the Fukushima disaster.

To add insult to injury, our government wants to re-start those old nuclear plants no longer in operation and even has the intention to export their own construction of nuclear power plants to other foreign countries.

To host the Olympic Games in a country still declared as being under nuclear emergency is absurd, and whoever participates take risk of being exposed, on the other hand; he or she is an accomplice to criminal conduct and is guilty of silence and denial.

845-002-0

註釈

・P34［＊1］事故から３ヶ月後に開かれた原子力安全に関するＩＡＥＡ（国際原子力機関）閣僚会議において日本政府は「東京電力福島原子力発電所の事故について」というレポートを提出した。その全文は「首相官邸」ホームページで読むことができる。
――原子力災害対策本部「原子力安全に関するＩＡＥＡ閣僚会議に対する日本国政府の報告書－東京電力福島原子力発電所の事故について」平成23年6月
https://www.kantei.go.jp/jp/topics/2011/iaea_houkokusho.html

・P44［＊2］2019年4月8日時点での汚染水の総量は113万6400トン。これは乗客定員5400名以上の世界最大客船「シンフォニー・オブ・ザ・シーズ」５隻分に相当する容量である。

・P52［＊3］東京電力はそれから半年後の2017年7月になって、容器直下で10グレイ以下、その外部であるペデスタルの壁と格納容器の壁の間では70グレイあるいは80グレイと、１時間当たりの放射線量の測定値の評価を下方修正した。
――国際廃炉研究開発機構　東京電力「２号機格納容器内部調査～線量率確認結果について」2017年7月27日
http://irid.or.jp/wp-content/uploads/2017/07/20170728_2.pdf

・P68［＊4］原子力緊急事態宣言は、原子力施設で極めて重大な事故が発生したとき、原子力災害対策特別措置法に基づき内閣総理大臣が発出する。東京電力福島第一原子力発電事故により2011年3月11日午後7時3分に初めて発令され、現在も継続中である。

・P76［＊5］東日本大震災から8年、福島県における避難者数は4万1299人（岩手県3666人、宮城県2038人）、震災関連死者数は2250人（岩手県467人、宮城県928人）、震災関連の自殺者は104人（岩手県50人、宮城県57人）に及ぶ。

Notes

• P34(*1) Three months after the accident, the Japanese government submitted a report "The Accident at TEPCO's Fukushima Nuclear Power Stations" at International Atomic Energy Agency (IAEA) Ministerial Conference on Nuclear Safety. The full report is available on the website "Prime Minister of Japan and Cabinet".

"Report of Japanese Government to the IAEA Ministerial Conference on Nuclear Safety - The Accident at TEPCO's Fukushima Nuclear Power Stations June, 2011

https://japan.kantei.go.jp/kan/topics/201106/iaea_houkokusho_e.html

• P45(*2) As of April 8, 2019, the total amount of the contaminated water is 1,136,400 tonnes. This amount is equivalent to the capacity of five "Symphony of the Seas", one of the largest cruise ships in the world, which accommodates more than 5,400 passengers.

• P53(*3) Half a year later, in July 2017, TEPCO made a downward adjustment on the evaluation of the radioactive value per hour; less than 10 grays directly under the pressure vessel, and 70 or 80 grays in the space between the pedestal wall of the pressure vessel and the wall of the containment vessel.

IRID&TEPCO "Unit 2 Containment Vessel Internal Investigation – Dose Rates Confirmation Result" July 27, 2017

http://irid.or.jp/wp-content/uploads/2017/07/20170728_2.pdf

• P69(*4) When a serious accident occurs at a nuclear power facility, the prime minister issues a declaration of a nuclear emergency situation based on Act on Special Measures Concerning Nuclear Emergency Preparedness. This declaration was first issued due to TEPCO Fukushima Daiichi Nuclear Power Station Accident at 7:03 on March 11, 2011, which is still in effect.

• P77(*5) Eight years after Great East Japan Earthquake, the number of evacuees in Fukushima prefecture reached at 41,299 (467 in Iwate prefecture, 2083 in Miyagi prefecture); earthquake related deaths at 2,250 (467 in Iwate, 928 in Miyagi), and earthquake related suicides at 104 (50 in Iwate, 57 in Miyagi).

[ドイツ語／German]

Die Katstrophe von Fukushima und die Olympischen Spiele in Tōkyō 2020

Augen vor der Wahrheit zu verschließen, ist ein Verbrechen

Koide Hiroaki (ehemaliger Forscher am Institut für Reaktorforschung der Universität Kyōto[*1])

Am 11. März 2011 wurde das von der Tōkyō Electric Power Company (kurz: TEPCO) betriebene Kernkraftwerk Fukushima Daiichi von einem gewaltigen Erdbeben erschüttert und anschließend von einem Tsunami überrollt, was zu einem totalen Stromausfall im gesamten Werk führte.

Die Experten waren sich einig, dass ein totaler Stromausfall, "die wahrscheinlichste Ursache für eine Katastrophe in einem AKW" sein würde. Und wie vorhergesagt, schmolzen die Reaktoren[*2] im Kernkraftwerk Fukushima Daiichi und verteilten große Mengen an radioaktiven Substanzen in die Umgebung. Einem Bericht[*3] der japanischen Regierung an die Internationale Atomenergie Organisation (IAEA) zufolge wurden durch diesen Unfall[*4] insgesamt 1,5x1016 Becquerel (Bq) an radioaktiven Substanzen, an Cäsium die 168-fache Menge der über Hiroshima abgeworfenen Atombombe, in die Atmosphäre freigesetzt. Die Radioaktivität einer einzigen Hiroshima-Atombombe ist gewiss schon angsteinflößend genug, aber hier spricht die japanische Regierung von der 168-fachen Menge an freigesetzter Radioaktivität.
Durch den Unfall schmolzen die Reaktoren der Blöcke 1, 2 und 3, in denen sich insgesamt 7x1017 Becquerel an Cäsium 137 befunden haben, was wiederum ungefähr 8000 Hiroshima-Atombomben entspricht. Wenn nun radioaktive Substanzen in der Menge von 168 Hiroshima-Atombomben in die Atmosphäre freigesetzt wurden, und dann noch die ins Meer gespülte Menge hinzugenommen wird, kann man wohl davon ausgehen, dass bis heute etwa das Tausendfache der Hiroshima-Atombombe in die Umwelt emittiert ist.
Cäsium 137 ist eines der radioaktiven Spaltprodukte, die bei der Kernspaltung von Uran entstehen, es ist diejenige radioaktive Substanz, die nach der Dreifachkatastrophe von Fukushima wohl die größte Bedrohung für die Menschen darstellt. Denn der Großteil der zuvor im Reaktor enthaltenen radioaktiven Substanzen befindet sich noch immer in den zerstörten Reaktorgebäuden und anderswo in Fukushima Daiichi. Ließe man den Reaktorkern weiter schmelzen, würden weitere radioaktive Substanzen, in denen auch Cäsium 137 gebunden ist, in die Umwelt austreten. Um dies zu verhindern, wird der geschmolzene Reaktorkern, von dem niemand sagen kann, wo genau er sich eigentlich befindet, bis heute – also acht Jahre nach dem Unglück – ununterbrochen gekühlt. Folglich fallen Tag für Tag hunderte Tonnen radioaktiv kontaminiertes Wasser an, das es zu lagern gilt.

TEPCO hat auf dem Werksgelände fast 1000 solcher Tanks aufgestellt, in denen das kontaminierte Wasser gelagert wird, dessen Menge inzwischen 1.000.000 Tonnen[*5] übersteigt. Die Tankkapazitäten sind ebenso begrenzt, wie es die Fläche des Werksgeländes ist. Folglich wird TEPCO sich bald gezwungen sehen, dieses radioaktiv kontaminierte Wasser ins Meer abzuleiten[*6]. Selbstverständlich wäre es das Wichtigste, die geschmolzenen Reaktorkerne so zu behandeln, dass sie so sicher wie möglich sind. Doch leider weiß auch heute nach mehr als acht Jahren niemand, wo und in welchem Zustand sich die Reaktorkerne befinden. Denn die Orte des Geschehens sind schlicht nicht zugänglich. Bei einem Unfall in einem kohle-, erdgas- oder erdölbetriebenen Heizkraftwerk wäre die Sache einfach. Dort würde vermutlich einige Tage lang ein Feuer wüten, ist dieses aber unter Kontrolle, kann der Unfallort betreten werden. Alles wird untersucht, repariert und der Betrieb wieder aufgenommen. Ereignet sich aber ein Unfall in einem Atomkraftwerk, so ist jeder Mensch, der sich zum Unfallort begibt, dem Tode geweiht.
Die Regierung und TEPCO wollten stattdessen Roboter einsetzen, mussten aber feststellen, dass diese der radioaktiven Strahlung nicht standhielten. Und zwar deshalb, weil die Mikrochips, auf denen ihre Befehle programmiert sind, durch die Einwirkung der radioaktiven Strahlung umprogrammiert werden. Aus diesem Grunde sind bisher fast alle Roboter nicht zurückgekehrt.
Ende Januar 2017 gelang es TEPCO, eine ferngesteuerte Kamera – ähnlich wie bei einer Darmspiegelung – in jene Betonkonstruktion einzuführen, auf der der Reaktordruckbehälter wie auf einem Sockel

steht. In der stählernen Plattform direkt unter dem Reaktordruckbehälter, auf der sonst Arbeiten an diesem durchgeführt werden, fanden sie ein großes Loch, was ihnen die Gewissheit verschaffte, dass der Reaktorinhalt durch den Boden des Druckbehälters hindurchgeschmolzen war und sich nun noch weiter unten befinden müsse.
Doch wurde durch diese Erkundung noch etwas viel Wichtigeres klar. Wird der menschliche Körper einer effektiven Dosis von circa acht Sievert pro Stunde ausgesetzt, so stirbt er. Unterhalb des Druckbehälters betrug die effektive Dosis pro Stunde circa 20 Sievert – ein unvorstellbarer Wert! Doch auf dem Weg durch die Öffnung dorthin betrug die radioaktive Strahlenbelastung an zwei Stellen 530 bzw. 650 Sievert pro Stunde[*7]. Zudem wurden diese Werte nicht im zylindrischen Unterbau gemessen, sondern im Zwischenraum zwischen der Wand des Sockels und dem Reaktorsicherheitsbehälter.
TEPCO und der Staat hatten zunächst ein Szenario entworfen, demzufolge der geschmolzene Reaktorkern, wie bei einem gefüllten Hefekloß, im Inneren des Sockels verbleibt, „um ihn nach 30 bis 40 Jahren zu bergen und in Castoren aufzubewahren. Das bedeutet die finale Eindämmung des Unfalls". Tatsächlich jedoch ist der geschmolzene Kernbrennstoff ins Äußere des Sockels gedrungen und ist in der Luft herumgespritzt. Das zwang den Staat und TEPCO, diesen „Fahrplan" umzuschreiben, nun hieß es, man werde den Reaktorsicherheitsbehälter an der Seite öffnen, um die Brennstäbe so bergen zu können. Allerdings halte ich dieses Vorhaben für unmöglich, da es zu einer enormen Strahlenexposition der Arbeiter führen würde.

Von Anfang an war ich der Meinung, dass wie zuvor beim Reaktorunglück in Tschernobyl in der ehemaligen Sowjetunion nur ein alles umhüllender Sarkophag infrage kommt. Der um das AKW Tschernobyl gebaute Sarkophag ist nach 30 Jahren in die Jahre gekommen und wurde im November 2016 von einem zweiten, noch größeren Sarkophag überzogen. Dieser zweite Sarkophag soll 100 Jahre halten. Welche Mittel danach zur Verfügung stehen, ist unklar.
Kein einziger der heute lebenden Menschen kann ein Ende des Unfalls in Tschernobyl absehen. Das gilt umso mehr für die Katastrophe in Fukushima, die selbst nach dem Tod aller heute Lebenden noch keinen Abschluss gefunden haben wird. Angenommen es gelänge, den geschmolzenen Reaktorkern in Behältern aufzubewahren, so würde damit ja die radioaktive Strahlung nicht verschwinden. Es müssten fortan diese Behältnisse für hunderttausende und Millionen von Jahren sicher verwahrt werden.
Auch in der Umgebung des Kraftwerks ist die unfassbare Tragödie längst nicht ausgestanden.
Als am Tag der Katastrophe der nukleare Ausnahmezustand[*8] ausgerufen wurde, wies man die Zwangsevakuierungen zunächst für einen Umkreis von drei Kilometern an, sie wurden dann auf 10 Kilometer und schließlich auf 20 Kilometer ausgeweitet, die Menschen flohen aus ihren Häusern, nur mit dem, was sie selbst mitnehmen konnten. Nutz- und Haustiere wurden zurückgelassen.
Im 40 bis 50 Kilometer von Fukushima Daiichi entfernten Dorf Iitate gab es darüber hinaus direkt nach der Katastrophe keinerlei Warnungen. Erst mehr als einen Monat nach der Katastrophe erging ein Evakuierungsbefehl aufgrund der dann doch festgestellten hohen Strahlenbelastung – das Dorf wurde vollständig evakuiert.

Was eigentlich nennen wir menschliches Glück?

Für die meisten Menschen bedeutet Glück wohl, gemeinsam mit ihrer Familie, ihren Freunden, ihren Nachbarn und ihren Geliebten in Ruhe ihrem Tageswerk nachgehen zu können, heute ebenso wie morgen und in den folgenden Tagen, immerfort.
Doch damit war es von einem Tag auf den anderen vorbei. Die Evakuierten wurden zunächst in Turnhallen und anderen größeren Räumen untergebracht, bevor sie dann zu zweit in gerade einmal acht m^2 kleinen Notunterkünften unterkamen, um dann wiederum in sogenannten „Katastrophen-Wiederaufbau-Häusern" oder in von lokalen öffentlichen Körperschaften angemieteten und den Evakuierten vorübergehend zur Verfügung gestellten Wohnungen einquartiert zu werden. Familien, die bis dahin zusammengelebt hatten, wurden nun getrennt. Ihr Alltag wurde zerstört, und immer wieder setzen Leute aus tiefer Verzweiflung ihrem Leben ein Ende[*9].
Doch nicht nur das. Auch außerhalb der wegen hohen Strahlenbelastung zwangsevakuierten Gebiete

entstanden weithin kontaminierte Regionen, die eigentlich zu sogenannten „Kontrollbereichen" hätten erklärt werden müssen. Es sind dies Orte, die nur von Erwachsenen betreten werden dürfen, di dafür bezahlt werden, dass sie sich der radioaktiven Strahlung aussetzen, also von beruflich strahlenexponierten Beschäftigten. Allerdings ist es selbst diesen beruflich strahlenexponierten Beschäftigte untersagt, innerhalb der Kontrollbereiche zu trinken oder gar zu essen – vom Schlafen ganz zu schweigen. Und weil es in Kontrollbereichen keine Toiletten gibt, kann man dort nicht einmal seine Notdurf verrichten. Doch hat der Staat unter dem Vorwand des akuten Notstandes die geltenden Gesetze auße Kraft gesetzt und Millionen Menschen dazu verdammt, in den kontaminierten Gebieten leben zu müssen.

Diese im Stich gelassenen Menschen, unter denen auch kleine Kinder sind, trinken, essen und schlafer dort. Und natürlich sind sie es, die sich mit den Gefahren der radioaktiven Strahlung herumzuschlage haben. Diese im Stich Gelassenen plagt die Angst. Um der Strahlenexposition zu entkommen, haber einige ihre Arbeit aufgegeben und sind mit der gesamten Familie geflohen. Andere wiederum wollter wenigsten ihre Kinder schützen, weshalb die Väter in den kontaminierten Regionen zurückblieber um zu arbeiten, und nur die Mütter mit den Kindern flüchteten. Das bedeutet jedoch, dass ihr Alltag ihre Familie zerstört wird. Bleiben sie in den kontaminierten Regionen, leiden ihre Körper, fliehen sie bricht ihnen das Herz.

Diese im Stich gelassenen Menschen leben nun schon seit mehr als acht Jahren tagein tagaus unter dieser Qual.

Dessen ungeachtet, hat der Staat alle diejenigen, die evakuiert worden oder freiwillig geflohen waren, angewiesen, ab März 2017 in die kontaminierten Regionen zurückzukehren, in denen die jährliche Strahlendosis unter 20 Millisievert liegt. Auch die – wie unzureichend auch immer – gezahlten Entschädigungen für verlorengegangen Wohnraum wurden eingestellt. Dies hat zur Folge, dass es Leute geben wird, denen nichts Anderes übrigbleibt, als in die kontaminierten Regionen zurückzukehren.

Gegenwärtig, so heißt es, werde dem Wiederaufbau in Fukushima höchste Priorität eingeräumt. Leben Menschen gezwungenermaßen dort, bleibt ihnen nur den Wiederaufbau zu erstreben. Es ist schlicht unmöglich tagtäglich in Angst zu leben. Sie möchten nicht jeden Tag an die Radioaktivität erinnert werden, und Radioaktivität – Segen oder Fluch? – ist unsichtbar. Staat und lokale Kommunen tragen zu ihrem tatsächlichen Vergessen bei. Daher werden all jene, die die Kontamination und die Angst thematisieren, als Störenfriede, die dem Wiederaufbau im Wege zu stehen, in Misskredit gebracht.

Eine jährliche Strahlendosis von 20 Millisievert entspricht der staatlich festgesetzten zulässigen Maximaldosis für beruflich strahlenexponierte Beschäftigte, wie sie einst auch für mich galt. Es ist einfach nicht hinnehmbar, diese Strahlendosis jetzt für Menschen zuzulassen, die aus dieser Exposition keinerlei Gewinn ziehen können. Zumal für Babys und Kinder, die durch die Strahlung besonders gefährdet sind und die an der nuklearen Amokfahrt Japans und am Fukushima-Unfall keinerlei Schuld tragen. Es darf einfach nicht sein, die Grenzwerte für beruflich strahlenexponierte Beschäftigte auch auf diese Menschen auszuweiten.

Doch behauptet der japanische Staat, dass „momentan der atomare Ausnahmezustand herrscht und es folglich nicht anders geht". Einen Tag, eine Woche, einen Monat oder, wenn es die Umstände erfordern, auch mal ein Jahr lang Ausnahmezustand, das ist irgendwie noch zu verstehen. Tatsächlich aber ist der verkündete „atomare Ausnahmezustand" auch nach acht Jahren noch immer nicht aufgehoben worden. Der Staat will den Unfall in Fukushima ganz bewusst vergessen machen, und weil auch die Massenmedien schweigen, vergisst der Großteil des Volkes, dass der „atomare Ausnahmezustand" nicht aufgehoben ist und gegen eigentlich geltende Gesetze verstoßen wird.

Der Hauptübeltäter der radioaktiven Substanzen, die die Umwelt verseuchen, ist Cäsium 137, dessen Halbwertszeit bei 30 Jahren liegt. Erst nach 100 Jahren ist es bis auf 10% gesunken. Und so wird sich dieses Land Japan auch in 100 Jahren noch im „atomaren Ausnahmezustand" befinden.

Die Olympischen Spiele wurden immer schon für die Beförderung von nationalem Prestige instrumentalisiert. In den letzten Jahren wurden sie zum Spielball einer enorm verschwenderischen Gesellschaft, in der Großprojekte gebaut und gleich wieder abgerissen werden, sowie von Unternehmenskonglomer-

aten mit Baufirmen im Zentrum, die daraus gigantische Profite schlagen.
In der gegenwärtigen Situation aber wäre es am wichtigsten, dass das Land alle Kräfte bündelt und darauf hinarbeitet, die„ Erklärung des nuklearen Ausnahmezustandes“ so rasch wie möglich aufzuheben. Den Menschen, die immer noch unter den Folgen der Katastrophe von Fukushima leiden, zur Seite zu stehen und ihnen zu helfen, sollte als Aufgabe oberste Priorität genießen. Wenigstens aber müssten die unschuldigen Kinder davor geschützt werden, radioaktiver Strahlung ausgesetzt zu sein.
Dessen ungeachtet hält der Staat an der Wichtigkeit der Olympischen Spiele fest.
Je tiefer die Krise im Inneren, desto größer die Anstrengungen der Mächtigen, von dieser abzulenken. Um Fukushima vergessen zu machen, werden die Massenmedien das Olympia-Fieber mehr und mehr entfachen, und es wird der Moment kommen, da jene, die gegen die Olympischen Spiele protestieren, als Verräter bezeichnet werden.
So war es auch im letzten Krieg.
Die Massenmedien brachten nur die Verordnungen des Kaiserlichen Hauptquartiers, und fast das gesamte Volk wirkte am Krieg mit. Je mehr man sich für einen hervorragenden Japaner hielt, desto eher verurteilte man einen Kriegsgegner in der Nachbarschaft als Verräter und schaltete ihn aus. In einem Land aber, das unschuldige Menschen im Stich lässt und „die Olympischen Spiele für wichtiger“ hält, ist es mir eine Freude zum Verräter zu werden.

Die Fukushima-Katastrophe wird fortdauern, gezählt in Jahrhunderten, einhergehend mit riesigen Tragödien.
Während die Täter, die Verursacher des Unfalls – TEPCO, Leute aus der Regierung, Wissenschaftler, Massenmedien – scheele Seitenblicke auf die enorm vielen Opfer werfen, hat kein einziger von ihnen Verantwortung übernommen. Niemand wurde verurteilt. Und als wäre das nicht schon genug, wollen sie nun stillstehende Kernkraftwerke wieder anfahren und diese Technik ins Ausland exportieren.

Olympische Spiele, die in einem Land im atomaren Ausnahmezustand veranstaltet werden.
All jene Länder und Menschen, die daran teilnehmen, setzen sich einerseits natürlich der Gefahr der Strahlenexposition aus, zum anderen aber machen sie sich zu Komplizen an den Verbrechen in diesem Land.

[*1] Seit März 2018 heißt diese Einrichtung Kyōto University Institute for Integrated Radiation and Nuclear Science. (Anm. d. Ü.)

[*2] Präzise gesprochen, schmolz nicht der Reaktor, sondern die mit radioaktiven Material bestückten Brennstäbe aufgrund ausbleibender Kühlung. (Anm. d. Ü.)

[*3] Der Bericht “Über den Unfall im von TEPCO betriebenen Kernkraftwerk Fukushima“ wurde drei Monate nach der Katastrophe von der japanischen Regierung im Rahmen der Ministerkonferenz der Internationalen Atomenergie Organisation (IAEA), die sich mit der Sicherheit von Kernenergie beschäftigte, eingereicht. Der vollständige Text ist über die Homepage des Premierminister-Amtssitzes einsehbar.
Quelle: „Bericht der japanischen Regierung an die Ministerkonferenz der IAEA zum Thema ‚Nukleare Sicherheit‘ – Zum Unfall im von TEPCO betriebenen Kernkraftwerk Fukushima“ (Juni 2011)
Japanisch: https://www.kantei.go.jp/jp/topics/2011/iaea_houkokusho.html
Englisch: https://japan.kantei.go.jp/kan/topics/201106/iaea_houkokusho_e.html

[*4] Sicher kann die Katastrophe auch als Unfall betrachtet werden, tatsächlich ereigneten sich jedoch mehrere Explosionen, weshalb folglich von einer Katastrophe zu sprechen ist, die aus unterschiedlichen Unfällen resultierte. (Anm. d. Ü.)

[*5] Am 8. April 2018 waren es 1.136.400 Tonnen Wasser. Dies entspricht dem fünffachen Fassungsvermögen des mehr als 5400 Menschen fassenden weltweit größten Passagierschiffs „Symphony of the Seas“.

[*6] Zu erwähnen ist, dass das Wasser nicht ungefiltert ins Meer abgelassen werden soll, sondern bereits Teile der radioaktiven Substanzen herausgefiltert werden. (Anm. d Ü.)

[*7] TEPCO berichtigte diese Werte ein halbes Jahr später, im Juli 2017 auf unter 10 Sievert pro Stunde für den Bereich direkt unter dem Reaktordruckbehälter und auf 70 bzw. 80 Sievert pro Stunde für den Bereich zwischen der Wand des

Reaktorsicherheitsbehälters und der Wand des darum liegenden Sockels.

Quelle: Bericht von TEPCO„ Untersuchung des Inneren des Reaktorsicherheitsbehälters in Reaktor 2 – Ergebnisse der Auswertung der Dosisrate" auf der Homepage des Internationelen Forschungsinstituts für den Nuklearen Rückbau (International Research Institute for Nuclear Decommisioning) vom 27.07.2017

http://irid.or.jp/wp-content/uploads/2017/07/20170728_2.pdf

[*8] Die „Erklärung des nuklearen Ausnahmezustands" wird vom Premierminister, im Falle eines außerordentlich großen Unfalls in einer kerntechnischen Anlage, auf der Grundlage des Gesetzes über Sondermaßnahmen im Fall von nuklearen Katastrophen ausgerufen. Erstmalig am 11. März 2011 um 17:03 Uhr aufgrund des Unfalls im von TEPCO betriebenen Atomkraftwerk Fukushima Daiichi ausgerufen, ist sie bis heute in Kraft.

[*9] Acht Jahre nach der Großen Erdbebenkatastrophe in Nord-Ost-Japan beträgt die Zahl der Evakuierten in der Präfektur Fukushima 41299 (in der Präfektur Iwate sind es 3666 Menschen und in der Präfektur Miyagi bis heute 2083 Menschen). Die Zahl der Toten, die im Zusammenhang mit der Katastrophe stehen, aber nicht direkt durch dies ums Leben kamen, beträgt in der Präfektur Fukushima 2250 Menschen, während es in der Präfektur Iwate 467 Menschen und in der Präfektur Miyagi 928 Menschen sind. Die Zahl der Selbstmorde, die als direkte Folge der Katastrophe zu betrachten sind, beläuft sich in der Präfektur Fukushima auf 104. In der Präfektur Iwate sind 50 und in der Präfektur Miyagi sind 57 Suizide zu beklagen.

Übersetzung ins Deutsche:
Prof. Dr. Steffi Richter, Felix Jawinski (Wissenschaftlicher Mitarbeiter)
Ostasiatisches Institut - Japanologie -
Universität Leipzig

[フランス語／French]

L'accident de Fukushima et les Jeux Olympiques de Tokyo

—C'est un crime de détourner le regard de la vérité.

Hiroaki Koide (ancien professeur adjoint à l›Institut de recherche nucléaire de l›Université de Kyoto)

Le 11 mars 2011, la centrale nucléaire de Fukushima Daiichi a été frappée par un énorme tremblement de terre et un tsunami, entraînant une panne de courant générale.

Les experts avaient conclu que la panne générale de courant serait « l'une des causes les plus probables de catastrophe nucléaire ». Comme prévu, les réacteurs nucléaires de la centrale nucléaire de Fukushima Daiichi sont fondus, répandant une grande quantité de matières radioactives dans l'environnement.
Selon un rapport soumis par le gouvernement japonais à l'Agence Internationale de l'Energie Atomique (AIEA)[*1], cet accident a entraîné la libération atmosphérique de 1.5×10 puissance 16 becquerels de césium 137, soit l'équivalent de 168 bombes atomiques d'Hiroshima. La radioactivité d'une bombe atomique d'Hiroshima est déjà terriblement effrayante, mais le gouvernement japonais a déclaré que la quantité de la radioactivité dispersée dans l'atmosphère était même 168 fois supérieure à celle-ci.
Cet accident a provoqué la fusion des cœurs des réacteurs 1, 2 et 3. Au total, 7×10 puissance 17 becquerels de césium 137, soit environ 8000 bombes atomiques d'Hiroshima , existait dans les cœurs. Parmi celles-ci, la quantité rejetée dans l'atmosphère équivalait à 168 bombes atomiques d'Hiroshima, et on peut estimer que la quantité de césium 137 rejetée dans l'environnement jusqu'à présent, y compris la quantité rejetée dans la mer, serait équivalente à 1000 bombes atomiques de Hiroshima.
Le césium 137 est l'un des isotopes fissiles produit de la fission nucléaire à partir d'uranium ; c'est un radioisotope qui représentait la plus grande menace pour l'humanité lors de l'accident de Fukushima.
En d'autres termes, une grande partie des radioisotopes des cœurs des réacteurs existent encore dans les constructions détruites de Fukushima Daiichi. Les radioisotopes contenant du césium 137 seront à nouveau rejetées dans l'environnement si les cœurs des réacteurs se détruisent davantage. Pour éviter que cela ne se produise, même après plus de huit ans depuis l'accident, ils versent sans cesse de l'eau vers les probables cœurs en fusion. Probable car il est toujours incertain l'emplacement de ces dernières.
Et de ce fait, des centaines de tonnes d'eau contaminée de radioactive continuent à être stockées chaque jour.

TEPCO a construit près de 1000 réservoirs sur le site pour stocker de l'eau contaminée, mais le montant total a déjà dépassé un million de tonnes.
L'espace du site est limité, donc le nombre de réservoirs qui peuvent être construits est limité. Bientôt, TEPCO sera obligée de jeter l'eau contaminée dans la mer[*2].
Bien sûr, le plus important c'est de ramener les cœurs fondus à un état plus stable, mais même après plus de huit ans, ils ignorent l'emplacement et l'état de cœurs parce qu'il n'est pas possible de pénétrer dans le site.
Si la centrale qui a provoqué l'accident était une centrale thermique, la maîtrise d'après accident était plus facile. Il pourrait y avoir une incendie pendant plusieurs jours, mais une fois éteint, on pourrait y pénétrer et voir à l'intérieur du site. Il deviendrait possible d'enquêter sur la condition de l'accident, de restaurer et de réactiver la centrale.
Cependant, si l'accident se produisait dans une centrale nucléaire, les gens mourraient s'ils se rendaient sur le site de l'accident.
Le gouvernement et TEPCO ont essayé d'envoyer des robots à la place, mais les robots sont vulnérables aux radiations, car si la puce de circuit intégré est exposée à des radiations dans lesquelles la commande est écrite, la commande elle-même peut être modifiée. Par conséquent, presque tous les robots envoyés jusqu'ici n'ont pas pu revenir.
Fin janvier 2017, TEPCO a inséré une caméra de contrôle à distance, assez similaire à celle d'un gastroscope, dans le socle en béton (Piédestal) sur lequel la cuve sous pression du réacteur était montée. Il s'est avéré qu'il y avait un grand trou dans la plateforme de travail en acier juste dessous de la cuve sous

pression, et que le cœur qui tombait du bas de la cuve s'est effondré davantage.
Mais encore cette recherche a révélé quelque chose de plus grave.
Les humains mourront sans exception s'ils sont exposés à 8 Sievert (Sv) sur tout le corps. La dose de rayonnement juste dessous de la cuve sous pression est de 20 Sv par heure, ce qui est déjà une dose trop élevée. Cependant, avant d'y arriver, un rayonnement de 530 ou 650 Sv a été mesuré[*3]. Et l'endroit où cette dose élevée a été mesurée n'était pas à l'intérieur du Piédestal cylindrique, mais entre le mur de Piédestal et le mur du conteneur de confinement.
TEPCO et le gouvernement ont voulu écrire un scénario selon lequel les cœurs fondus seraient empilés comme des boulettes à l'intérieur de Piédestal, et que « Dans 30 à 40 ans, ils peuvent récupérer ces boulettes et les sceller dans les conteneurs. Ils concluent ainsi comme la fin de l'accident. »
Mais En réalité, les combustibles nucléaires en fusion sont coulés hors de Piédestal et se sont dispersés. Inévitablement, le gouvernement et TEPCO étaient obligés de réécrire la « feuille de route » et ont annoncé qu'ils allaient percer un trou dans le flanc du conteneur et les saisir. Toutefois, si un tel travail est effectué, la dose de rayonnement sur les travailleurs sera énorme. Ce plan n'est pas réalisable.

Depuis le début, j'ai insisté que la seule solution est de le sceller dans le sarcophage, comme ce fut le cas lors de l'accident de la centrale nucléaire de Tchernobyl dans l'ex-Union soviétique.
Après 30 ans, ce sarcophage est complètement usé et en novembre 2016 ils ont le recouvert par un autre sarcophage encore plus gigantesque. La longévité de ce deuxième serait de 100 ans. Après cela, on ne sait quelle mesure pourrait être prise.
Personne de vivant aujourd'hui ne verra la restauration de l'accident de Tchernobyl.
De même après que toutes les personnes vivantes aujourd'hui soient mortes, le processus de restauration de l'accident de Fukushima ne sera pas complet.
Même si nous trouvons un moyen de sceller les cœurs en fusion dans des conteneurs, cela ne signifie pas que le rayonnement disparaît pour le bon. Pendant des centaines de milliers d'années ou même un million d'années, les conteneurs doivent être gardés en sécurité.
Même à l'heure actuelle, une tragédie extrême se poursuit dans l'environnement proche de la centrale.
Le jour de l'accident, une déclaration d'état d'urgence [*4] a été annoncée et la zone d›évacuation forcée est passée de 3 km à 10 km, puis à 20 km du site. Les gens ont quitté leurs domiciles avec le minimum nécessaire. Les bétails et les animaux de compagnie ont été délaissés.
Le village de Iijima, situé à une distance de 40 à 50 km de la centrale nucléaire de Fukushima Daiichi, n'a reçu aucun avertissement ni instruction d'évacuation immédiate après l'accident, mais plus d'un mois après , tout le village avait reçu l'instruction d'évacuation, car l'endroit s'est avéré extrêmement contaminé.

Que signifie le bonheur pour les gens ordinaires?

J'imagine que c'est la continuité des jours paisibles avec la famille, les amis, les voisins et les êtres chers sans soucier pour demain, après-demain et pour toujours.
Mais un jour ce bonheur a été arraché d'un coup.
Les personnes évacuées ont d'abord été envoyées dans des abris tels que des gymnases, puis dans de minuscules maisons temporaires dotées de quatre tatamis et demi (environ 7,4 mètres carrés) pour deux personnes, ou dans des logements de secours et autres logements temporaires. Pendant ce temps, les familles qui vivaient ensemble jusque-là ont été dispersées. Il y a eu beaucoup de gens dont la vie a été totalement détruite et qui se sont suicidés par désespoir, et sera dure jusqu'à maintenant[*5].
Ce n'est pas tout. En dehors de la zone où l'évacuation forcée était ordonnée en raison d'une contamination extrême, un grand nombre de zones contaminées auraient dû être classées comme la « zone de contrôle de rayonnement ».
La «zone de contrôle de rayonnement» est une zone où seuls les adultes salariés qui traitent le rayonnement et les opérateurs spécialisés sont autorisés à entrer. Et ces travailleurs, une fois qu'ils entrent dans cette zone contrôlé de rayonnent, n'ont pas le droit ni de boire de l'eau ni de manger de la nourriture. Bien évidemment, il est également interdit d'y dormir. Il n'y a pas de toilettes dans la zone de contrôle

de rayonnement et ils ne peuvent pas faire leurs besoins. Cependant, au mépris des ce règlement conventionnel, le gouvernement, en disant qu'il s'agissait d'une urgence, a abandonné des millions de personnes dans la zone contaminée et les a forcées à y vivre.
Ces personnes abandonnées, y compris les bébés, boivent de l'eau, mangent, et dorment là. Naturellement, ils sont exposés aux risques de radiations. Tous ces gens-là doivent se sentir en danger. Il y a des personnes qui ont abandonné leur travail et qui se sont évacuées en famille afin d'éviter l'exposition aux radiations. Il y a aussi des familles qui, voulant protéger les enfants d'irradiation, ont décidé de garder le père dans la zone contaminée pour le travail et d'évacuer les enfants et la mère. Ainsi certaines de leurs vies ou des familles se sont effondrées. Rester dans une zone contaminée endommagera leurs santés et l'évacuation brisera leurs cœurs.
Pendant plus de huit ans depuis l'accident, ces personnes abandonnées vivent dans la détresse chaque jour.
Même alors, en mars 2017, le gouvernement a ordonné aux personnes qui ont été évacuées ou qui ont évacué volontairement de retourner dans leur région d'origine si la contamination ne dépassait pas 20 milli-sieverts (mSv) par an. Le gouvernement a également supprimé l'indemnité de logement qui les aidait tout de même à survivre. En coupant le soutien financier, certaines personnes n'auront d'autre choix que de retourner dans la zone contaminée.

À Fukushima, la reconstruction est considérée comme plus importante que toute autre chose.
Il n'y a aucun doute, pour les habitants, qu'ils espèrent que la reconstruction rapide, surtout s'ils n'ont pas d'autre choix que d'y vivre. En outre, personne ne peut vivre avec la peur en permanence. On voudrait oublier qu'il y a de la contamination et que, pour le meilleur ou pour le pire, le rayonnement est invisible. Le gouvernement et l'administration locale tentent de leur faire oublier ce problème de manière proactive. S'ils parlent de la contamination et de leur anxiété, ils seront accusés d'avoir entravé la reconstruction.
La dose d'exposition de 20 milli-sieverts par année était la limite d'exposition que le gouvernement avait initialement définie pour les « travailleurs sous rayonnement », ce qui était mon cas. Il est impardonnable d'appliquer la même mesure à ceux qui ne reçoivent aucune compensation par l'exposition aux rayonnements. De plus, les bébés et les enfants sont sensibles à l'exposition aux radiations et ils ne sont pas responsables du désastre de l'accident de Fukushima provoqué par le plan politique du nucléaire japonais qui est devenu incontrôlable. Appliquer le même standard de travailleurs sous rayonnement à la population générale est quelque chose que l'on ne devrait jamais faire.
Cependant, le gouvernement japonais a déclaré: « Il n'y a pas d'autre moyen, car nous sommes en situation d'urgence nucléaire ».
Si la situation d'urgence durait qu'une journée, une semaine, un mois ou un an au plus, je serais en mesure de mieux comprendre ce protocole, mais la « déclaration d'urgence nucléaire » n'est pas encore levée, même huit ans après l'accident.
Le gouvernement tente délibérément de faire oublier l'accident de Fukushima et les médias gardent le silence à ce sujet. C'est pourquoi les plupart de peuples ignorent le fait que la « déclaration d'urgence nucléaire » n'est pas encore levée, ainsi que le fait que les règlements conventionnels sont blâmés.
La principale contamination radioactive qui pollue l'environnement provient du césium 137, dont la demi-vie est de 30 ans. Même après 100 ans, il n'est réduit qu'à un dixième.
En fait, ce pays, le Japon, sera soumis à la « déclaration d'urgence nucléaire » même après 100 ans.

Les Jeux olympiques ont été toujours utilisés pour renforcer le prestige national.
Ces dernières années, les Jeux Olympiques ont été la proie d'une société de gaspillage qui construit et détruit les bâtiments. Le grand nombre d'entreprises, notamment les entreprises de construction, tiraient donc d'énormes profits.
Cependant, la mission la plus importante et la plus urgente devrait être de faire lever la « déclaration d'urgence nucléaire » en ramassant toute puissance nationale. Soulager ceux qui ont souffert de l'accident de Fukushima devrait être notre priorité absolue. À tout le moins, nous devons protéger les enfants innocents contre l'exposition aux rayonnements.

Néanmoins, le gouvernement affirme que les Jeux olympiques sont importants.
Plus nous avons de problèmes à l'intérieur du pays, plus les autorités tentent de détourner l'attention des gens. Et pour faire oublier Fukushima, les médias vont accélérer davantage le boom olympique. Bientôt, il arrivera un temps où quiconque s'oppose aux Jeux Olympiques sera considéré comme antipatriotique.
La même chose était arrivée pendant la guerre précédente.
Les médias ont simplement publié l'annoncé du siège principal et presque toute la population a coopéré à la guerre. Les personnes qui se considéraient comme d'excellentes nationalistes japonaises ont accusé leurs voisins opposés à la guerre, affirmant qu'ils étaient antipatriotiques, et les ont opprimés. Cependant, si notre pays donne la priorité aux Jeux Olympiques plutôt qu'au sauvetage de ces innocents, je serais heureux d'être traité d'antipatriotique.

L'accident de Fukushima continuera à nous toucher avec les énormes tragédies encore des centaines d'années à venir.
TEPCO, les responsables gouvernementaux, universitaires et les médias, qui sont les auteurs de cet accident, n'assument aucune responsabilité, même lorsqu'ils sont bien conscients d'un grand nombre de victimes existant. Ils ne sont pas traduits en justice non plus. Ils en profitent et tentent de réactiver les centrales nucléaires qui sont actuellement fermées et même cherchent à exporter l'industrie nucléaire à l'étranger.

Les Jeux Olympiques de Tokyo se dérouleront dans un pays sous la déclaration d'urgence nucléaire.
Les pays participants et leurs populations seront bien sûr exposés au risque d'exposition aux rayonnements. En outre, ils joueraient un rôle dans le soutien des crimes commis par ce pays.

[*1] Trois mois après l'accident, le gouvernement japonais a présenté à la conférence ministérielle de l'AIEA (Agence internationale de l'énergie atomique) sur la sûreté nucléaire, un rapport intitulé « L'accident survenu aux centrales nucléaires de Fukushima de TEPCO ». Le texte intégral peut être lu sur la page d'accueil du "Cabinet du Premier ministre".

Rapport du gouvernement japonais à la Conférence ministérielle de l'AIEA sur la sûreté nucléaire - L'accident survenu aux centrales nucléaires de Fukushima de TEPCO - publié en juin 2011

[*2] Le 8 avril 2019, la quantité totale d'eau contaminée était de 1 136 400 tonnes. Cette capacité équivaut à cinq navires « Symphony of the Seas », le plus grand navire à passagers au monde d'une capacité de 5 400 passagers ou plus.

[*3] Six mois plus tard, en juillet 2017, TEPCO a baissé l'évaluation de la valeur mesurée de la dose de rayonnement par heure à moins de 10 grays pour la partie située juste dessous du conteneur et à 70 ou 80 grays pour la partie située entre le mur extérieur de Piédestal et le mur du conteneur de confinement.

L'institut international de recherche sur le déclassement des installations nucléaires et TEPCO « Résultats de l'enquête interne de la centrale nucléaire de Fukushima Daiichi 2 - Résultats de l'enquête interne sur les conteneurs de confinement ». 27 juillet 2017

[*4] La Déclaration d'urgence nucléaire est promulguée par le Premier ministre en vertu de la Loi sur les mesures spéciales concernant la préparation aux urgences nucléaires lorsqu'un accident critique se produit dans une installation nucléaire. Elle a été émise pour la première fois à 19h03 le 11 mars 2011 en raison de l'accident de la centrale nucléaire TEPCO Fukushima Daiichi et elle est toujours en cours.

[*5] Huit ans après le grand désastre sismique à l'Est du Japon, le nombre d'évacués dans la préfecture de Fukushima est de 41 299 (3 666 dans la préfecture d'Iwate et 2 083 dans la préfecture de Miyagi), et le nombre de morts liés au séisme est de 2 250 (467 à Iwate et 928 à Miyagi) Le nombre de suicides liés au séisme est de 104 (50 à Iwate et 57 à Miyagi).

[スペイン語／Spanish]

El accidente de Fukushima y las Olimpíadas de Tokio

—Es un crimen apartar la vista de la verdad

Koide Hiroaki (ex profesor asistente del Instituto de Investigaciones de Reactores Nucleares de la Universidad de Kioto)

El 11 de marzo de 2011, la central nuclear Fukushima Número 1 de la Compañía Eléctrica de Tokio (TEPCO) fue azotada por un enorme terremoto y por un tsunami, provocando un apagón total de las instalaciones.

El apagón fue, según acordaron los expertos, "la causa más probable de la sucesiva catástrofe nuclear". Esto habría causado el derretimiento del reactor nuclear y la diseminación de vastas cantidades de material radioactivo en los alrededores. De acuerdo a un reporte emitido por el gobierno de Japón ante el Organismo Internacional de Energía Atómica (OIEA)[*1], el accidente provocó la fuga a la atmósfera de 1.5 x 1016 becquerel y de una cantidad de cesio-137 equivalente a 168 bombas atómicas de Hiroshima. La radiación de una sola bomba atómica es más que atemorizante, pero el gobierno confirmó que la radicación diseminada en el aire equivalía a 168 de ellas.

Todavía más, el accidente implicó un derretimiento de los reactores 1, 2 y 3, en cuyos núcleos había un total de 7 x 1017 becquerel y un promedio de cesio-137 igual a 8000 veces lo que sería una bomba atómica como la de Hiroshima. Si efectivamente 168 partes se liberaron a la atmósfera, y la cantidad que se había liberado al mar, puede estipularse que el total de material radioactivo fugado al medio ambiente hasta la fecha es de alrededor de 1000 veces de lo que sería una bomba atómica.

El cesio-137 es un producto nuclear que se genera a partir de la fisión de uranio y es la mayor amenaza para las personas que dejó como consecuencia el accidente de Fukushima. Una gran parte de este material radioactivo que existía en el reactor original continúa existiendo hoy en los restos de los edificios dañados de la central. Esto es, si vuelve a ocurrir otro derretimiento del núcleo del reactor, sólo con dichos restos volverían a liberarse a la atmósfera materiales radioactivos con cesio-137. A fin de prevenir sus efectos, se sigue arrojando agua fría al lugar en donde podría haber estado el núcleo ahora derretido, incluso hoy que han pasado más de ocho años desde el accidente. El resultado de esto es, sin embargo, que cientos de toneladas de agua contaminada con materiales radioactivos se continúan acumulando cada día.

TEPCO construyó unos 1000 tanques en las instalaciones para conservar el agua contaminada, pero la cantidad total de la misma ya ha excedido el millón de toneladas[*2]. El espacio en las instalaciones es limitado y por lo tanto también el número de tanques. En un futuro cercano, TEPCO no tendrá más opción que liberar el agua con material radioactivo al mar.

Desde ya, lo más importante es procurar que el núcleo derretido se mantenga en un estado de relativa estabilidad, pero ahora que han pasado ya ocho años del accidente es difícil saber en dónde se encuentra dicho núcleo o bien cuál es su condición. Y esto se debe a que no puede ingresarse al lugar.

Si la planta en que ocurrió el accidente hubiese sido una central termoeléctrica, todo habría sido más sencillo. El incendio habría continuado inicialmente por algunos días, pero luego se habría podido ingresar a las instalaciones luego de que se normalizara la situación. Habría sido posible examinar el estado del lugar tras el accidente, realizar las restauraciones necesarias y reactivar el reactor.

Sin embargo, el accidente ocurrió en una planta de energía nuclear. Y siendo ese el caso, las personas que ingresen al lugar del siniestro morirían.

El gobierno y TEPCO intentaron enviar robots para que ingresaran al área, pero estos son frágiles ante la exposición a la radiación. Esto se debe a que, si los rayos radioactivos alcanzan el chip IC en que están inscritas las órdenes al robot, las mismas pueden resultar alteradas. Es por esto que la mayoría de los robots enviados no han podido regresar hasta el momento.

A fines de enero de 2017, TEPCO puso una cámara a control remoto similar a un fibroscopio dentro del pedestal de cemento que sostenía el recipiente del reactor. Se descubrió un enorme agujero abierto en el andamio metálico de trabajo, directamente bajo el recipiente de presión; el núcleo se había derretido acarreando consigo la base de este último y cayendo todavía más al vacío.

Sin embargo, se descubrió una cosa todavía más importante en dicha inspección.

Una persona que se vea expuesta a 8 sievert de radiación de seguro morirá. La radiación bajo el recipiente del reactor era de unos 20 sievert por hora, una tremenda cantidad. Sin embargo, antes de alcanzar ese espacio, se registraron enormes cantidades de radiación, no en el interior del pedestal cilíndrico, sino en el espacio entre la pared del mismo y el contenedor. Dichas mediciones fueron de 530 y hasta 650 sievert de radiación[*3]. Originalmente, TEPCO y el gobierno afirmaron que el núcleo derretido se había acumulado dentro del pedestal como el relleno dentro de un dulce. "En 30 o 40 años todo el núcleo estará dentro del recipiente. Y ése será el fin de este accidente", aseguraron.

Sin embargo, la realidad comprobó que el combustible nuclear derretido se rebalsó del contenedor y se dispersó por otros espacios. Reacios ante esta situación, TEPCO y el gobierno empezaron a decir que reformularían sus planeamientos y que abrirían un agujero en el costado del contenedor para realizar extracciones. Sin embargo, de realizar dicha operación, los operadores se verían expuestos a una enorme cantidad de radiación. Sencillamente, no podía hacerse.

Desde un principio he afirmado que no nos queda más que sellar el lugar como si fuese una tumba, de igual modo que se hizo tras el accidente de la central nuclear de Chernóbil en la antigua URSS.

Hoy, 30 años después, dicha tumba comenzó a resquebrajarse y fue recubierta en noviembre de 2016 por una segunda tumba, con una longevidad de 100 años, según se dijo. Luego de esto, desconocemos qué medidas será posible tomar.

Ni una sola persona que esté viva hoy podrá ver la solución a las consecuencias del accidente de Chernóbil.

De más está decir que las consecuencias del accidente de Fukushima no van desaparecer tras la muerte de todas aquellas personas que lo presenciaron.

Aún si fuese posible sellar el núcleo derretido dentro del contenedor, eso no significa que se hayan eliminado todos los materiales radioactivos también. Por cientos de miles o millones de años, los contenedores deberán mantenerse resguardados y bajo vigilancia.

Las inmediaciones de la planta nuclear también están siendo atravesadas por una terrible tragedia.

El día del accidente se pronunció una Declaración de Emergencia Nuclear[*4] y se difundieron órdenes de evacuación obligatoria en un radio de 3 kilómetros del área, luego de 10 kilómetros y finalmente en 20 kilómetros a la redonda. Las personas tuvieron que abandonar sus casas llevando tan sólo lo que podían llevar en sus manos. El ganado y las mascotas también fueron abandonados.

Asimismo, la aldea de Iitate, que está a 40 o 50 kilómetros de la Central y la cual no recibió ningún tipo de alarma o instrucción al momento del accidente, finalmente debió ser evacuada un mes después del siniestro por estar terriblemente contaminada.

Qué es lo que suele decirse sobre la felicidad de las personas?

Para muchos, la felicidad significa pasar un día tranquilo con familiares, amigos, vecinos y parejas, pero también continuar de ese modo mañana, pasado y los días siguientes.

Hasta que un día ocurre un quiebre abrupto.

Las personas evacuadas tras el accidente fueron transferidas a gimnasios e instalaciones usadas como refugios; luego, a albergues de unos pocos metros para dos personas, a instalaciones especiales para la recuperación ante catástrofes y a pensiones públicas. A lo largo de ese proceso, las familias cuyos miembros habían vivido juntos hasta ese momento empezaron a desmoronarse. Al día de hoy es incalculable el número de personas cuyas vidas cotidianas se vieron destruidas o que incluso se quitaron la vida a causa de la desesperación[*5].

Pero eso no es todo. Fuera del área en que se realizó la evacuación obligatoria debido a la contaminación extrema surgieron otras áreas contaminadas que debían de haberse catalogado como "Zona de Radiación Controlada".

Una "Zona de Radiación Controlada" es aquella dentro de la cual se permite el ingreso sólo a trabajadores mayores de edad que sean capaces de manipular materiales radioactivos y que reciban una remuneración por ello. Aunque estos trabajadores ingresen a la zona, sin embargo, tienen prohibido beber el agua y comer la comida que allí se encuentra. Por supuesto, dormir allí también está prohibido. En una Zona de Radiación Controlada no hay baños y tampoco se puede defecar. En el marco de la situación de emergencia, el gobierno ignoró las regulaciones previas y abandonó a millones de personas dentro de este tipo de zonas contaminadas, como si los estuviera forzando a vivir allí.

Las personas abandonadas, incluso los bebés, tuvieron que beber el agua y comer la comida de esas zonas, así como también se vieron obligados a dormir allí. Desde ya, esto acarreó un enorme peligro por la exposición a la radiación. Es posible afirmar que ninguna de las personas abandonadas estuvo realmente a salvo. A fin de evitar la exposición, algunas de esas personas abandonaron sus trabajos y evacuaron la zona junto a todos sus familiares. En otros casos, a fin de proteger a sus hijos, los padres se quedaron a trabajar en la zona mientras que los niños y las madres evacuaron el lugar. Cualquiera la decisión, la cotidianeidad y las familias de estas personas se derrumbaron. De quedarse en las áreas contaminadas, su integridad física resultaba puesta en riesgo. De abandonar el lugar, era su corazón el que se quebraba en pedazos.

Las personas abandonadas vivieron con angustia cada día de los últimos ocho años.

Y, aun así, en marzo de 2017 el gobierno ordenó a todas las personas evacuadas por ellos o voluntariamente a regresar en el plazo de un año a todas aquellas áreas que no excedieran los 20 mSv (millisievert) de radiación. También interrumpió el subsidio para viviendas con el que de alguna manera los ayudaba hasta entonces. El tener que cumplir con esa orden no les dejó a muchas personas otra opción más que regresar a una zona contaminada.

Hoy en día se considera que la reconstrucción en Fukushima es lo más importante de todo.

En tanto que la situación para muchas personas sea el tener que vivir allí, no es posible otra cosa más que rogar por una reconstrucción. Después de todo, la gente no puede vivir cada día presa del miedo. Y la radiación es, para bien o mal, invisible; podemos olvidar que siquiera existe. Además, el gobierno nacional y el local van a promover que olvidemos el asunto de una buena vez. Y así, cuando uno hable sobre contaminación, será acusado de obstaculizar la reconstrucción.

El límite de exposición permitido desde un comienzo por el gobierno para los "trabajadores de la radiación" (como lo era entonces yo) era de 20 mSv (millisievert) al año. Pero luego este parámetro fue usado para todos los afectados. Es difícil tolerar la existencia de personas que no reciben algún tipo de remuneración tras haber sido expuestas a esa cantidad de radiación. Los bebés y niños son sobre todo sensibles a esta última y no tienen ninguna responsabilidad, ni del derrame nuclear ni del accidente. De más está decir, asimismo, que no deberían usarse los parámetros de los "trabajadores de la radiación" a las personas comunes y corrientes.

Sin embargo, el gobierno japonés aseguró que "no queda más remedio puesto hemos tenido que declarar la emergencia nuclear". Podemos comprender que una situación de emergencia dure un día entero, una semana, un mes, un año en caso de ser un hecho lamentable. Pero esta "Emergencia Nuclear" se mantuvo incluso ocho años después del accidente.

El gobierno intenta deliberadamente hacer que la gente olvide el incidente de Fukushima. Y los medios de comunicación mantienen sus bocas cerradas. Pero dado que dicha "Emergencia Nuclear" no fue rescindida hasta el día de hoy, gran parte de la población está olvidando que la ley original sigue siendo violada.

El más peligroso de los materiales radioactivos que contaminan el medio ambiente es el cesio-137, cuyo período de semidesintegración es de 30 años. Sin embargo, aunque pasen 100 años, éste se reduce sólo en un 10%. Lo cierto es que este país al que llamamos Japón estará bajo "Emergencia Nuclear" de aquí a un siglo en adelante.

Las Olimpíadas fueron usadas en todas las épocas para exaltar el prestigio nacional.

En años recientes, una creciente sociedad de consumo que fabrica tantas como destruye, y los grupos económicos centrados en el campo de la construcción que tan enormes ganancias brinda, han hecho de las Olimpíadas su objetivo predilecto.

Sin embargo, lo más importante hoy en día debería ser trabajar en conjunto como país para eliminar la "Emergencia Nuclear" cuanto antes. El tema prioritario debería ser la asistencia a las personas que siguen sufriendo tras los efectos del accidente de Fukushima. Cuanto menos, deberíamos proteger a los niños inocentes de la exposición a la radiación.

Aun así, el gobierno asegura que las Olimpíadas son importantes.

Cuando más arraigada en el interior esté la crisis, tanto más querrán los poderosos hacernos apartar la vista de ella. Así, para hacernos olvidar de Fukushima, los medios de comunicación infundirán más y más la fiebre olímpica. Incluso vendrá un tiempo en que dirán que aquellos que se oponen a las Olimpíadas son enemigos de la patria.

Así fue como sucedió en los años de la guerra.

Los medios de comunicación difundían solo los anuncios del Cuartel General Imperial, logrando así que casi todos los ciudadanos colaboraran. Cuanto más se consideraba a sí misma una persona como japonés ejemplar, tanto más condenaba a sus vecinos que estaban en contra de la guerra como anti-patriotas, ignorando sus opiniones. Sin embargo, en un país que dice que "las Olimpíadas son importantes" a la vez que se deshace de personas inocentes, yo no dudaré un segundo en convertirme en un anti-patriota.

El accidente de Fukushima acarrea una gran tragedia que se extenderá en los próximos cien años.

Mientras miramos de costado a la gran cantidad de víctimas, no hay uno solo de los culpables del accidente, ya sea TEPCO, los funcionarios de gobierno, los académicos, los representantes de los medios de comunicación, que haya asumido la responsabilidad. Nadie ha sido tampoco castigado. Para ellos, lo bueno es que se estén reactivando las plantas nucleares que estaban fuera de funcionamiento y que se las esté exportando fuera del país.

Las Olimpíadas de Tokio se realizarán en un país que en que rige una "Emergencia Nuclear".

Por un lado, las personas que participen de las mismas correrán el peligro de estar expuestas a la radiación. Por el otro, serán cómplices de este crimen que persiste en el país.

[*1] Tres meses después del siniestro, el gobierno japonés presentó el reporte "Sobre el accidente de la central nuclear Fukushima Número 1 de la Compañía Eléctrica de Tokio (TEPCO)" ante la cúpula del Organismo Internacional de Energía Atómica (OIEA). El texto puede leerse en el sitio de Internet del Primer Ministro.

"Sobre el accidente de la central nuclear Fukushima Número 1 de la Compañía Eléctrica de Tokio (TEPCO): Reporte del gobierno de Japón a los miembros de la OIEA sobre las medidas de seguridad tomadas". Departamento de emergencias nucleares, junio de 2011.

https://www.kantei.go.jp/jp/topics/2011/iaea_houkokusho.html

[*2] La cantidad total de agua contaminada para el 8 de abril de 2019 es de 1,136,400 toneladas. Dicha cifra corresponde a cinco veces la capacidad del barco de pasajeros más grande del mundo, el Symphony of the Seas, capaz de llevar a más de 5400 pasajeros.

[*3] Medio año después de la inspección, en julio de 2017, TEPCO modificó esas cifras a 10 grays directamente bajo el reactor y a 70 u 80 grays en el espacio entre la pared del pedestal exterior y la del recipiente, lo que significa que las cifras de las mediciones de radiación por hora fueron deliberadamente rebajadas.

"Sobre los resultados de la cantidad de radiación tras la inspección del interior del recipiente del reactor Unidad Número 2". Organismo de Investigación y Desarrollo para el Desmantelamiento Nuclear, TEPCO, 27 de julio de 2017.

http://irid.or.jp/wp-content/uploads/2017/07/20170728_2.pdf

[*4] La Declaración de Emergencia Nuclear la dictamina el Primer Ministro luego de un accidente de enormes proporciones en una instalación nuclear y basándose en Ley de Medidas Especiales Relacionadas a las Emergencias Nucleares. La Declaración fue emitida a las 19:03 del 11 de marzo de 2011 en respuesta al accidente de la Central Nuclear Fukushima Número 1 de TEPCO. Aún sigue en vigencia.

[*5] Pasados ocho años de aquel gran terremoto, 41299 personas habían sido evacuadas dentro de la prefectura de Fukushima (3666 personas en la prefectura de Iwate y 2083 personas Miyagi), 2250 personas perdieron la vida durante el mismo (467 personas en la prefectura de Iwate y 928 personas en la prefectura de Miyagi) y 104 se suicidaron por motivos relacionados al evento.

【ロシア語／Russian】

Авария на АЭС Фукусима-1 и Олимпийские игры в Токио :

Закрывать глаза на правду – это преступление

Коидэ Хироаки (ранее доцент Научно-Исследовательского Института Ядерных Реакторов при Университете Киото)

11 марта 2011 года АЭС Фукусима-1, управляемая Токийской энергетической компанией, подверглась удару мощнейшего землетрясения и цунами, что привело к её полной остановке.

Специалисты сошлись во мнении, что полная остановка электростанции «является наиболее вероятной причиной возникновения разрушительной ядерной катастрофы». В соответствии с этим предположением, расплавление активной зоны реакторов АЭС Фукусима-1 вызвало выброс огромного количества радиоактивных веществ в окружающую среду.
Согласно докладу правительства Японии для МАГАТЭ[*1], в результате аварии в атмосферу было выброшен цезий-137 активностью 1.5×1016 Бк, что эквивалентно 168 ядерным взрывам в Хиросиме. Радиоактивный потенциал одного только взрыва в Хиросиме устрашает, а теперь японское правительство заявляет, что в атмосферу попало в 168 раз больше радиоактивных веществ.
Авария привела к расплавлению активной зоны первого, второго и третьего реакторов, в которых находилось достаточное для 8000 взрывов в Хиросиме количество цезия-137 общей активностью 7×1017 Бк. В атмосфере оказалось количество радиоактивных веществ, достаточное для 168 ядерных взрывов, а если добавить ещё то, что попало в море, то получится, что до сегодняшнего дня в окружающую среду было выброшено радиоактивных веществ в эквиваленте примерно 1000 взрывов в Хиросиме.
Цезий-137 – это один из продуктов ядерного распада, образующийся при распаде урана; он является самым опасным для человека радиоактивным веществом, выброшенным в окружающую среду вследствие аварии. А немалая часть радиоактивных веществ, которые были в реакторе, всё ещё находятся в разрушенных помещениях АЭС Фукусима-1. Более того, если произойдет расплавление ядра реактора, то в окружающую среду снова попадут радиоактивные вещества, в том числе цезий-137. Чтобы это предотвратить, даже сейчас, спустя более чем 8 лет, в реакторы постоянно закачивают воду. Из-за этого ежедневно образуются сотни тонн радиоактивной воды.

Токийская энергетическая компания установила на прилегающей территории 1000 цистерн для сбора зараженной воды, и на сегодняшний день её накоплено уже более миллиона тонн[*2].
Территория не бесконечна, есть и предел увеличению количества цистерн. В ближайшем будущем Токийской энергической компании придётся сбрасывать зараженную воду в море.
Разумеется, наиболее важно привести реакторы в более или менее безопасное состояние, но даже сейчас, спустя более чем 8 лет, неясно, где и в каком состоянии находится расплавившееся ядерное топливо. Всё потому, что нельзя попасть в зону аварии.
Если бы авария произошла на теплосиловой электростанции, всё было бы просто. Сначала, вероятно, несколько дней бы продолжался пожар, но, потушив его, можно было бы посетить место происшествия. Узнать, что вызвало аварию, восстановить электростанцию и запустить её снова.
Но если человек окажется на территории АЭС, на которой произошла авария, он умрёт.
Правительство и Токийская энергетическая компания решили взамен людей отправить туда роботов, но они неустойчивы к излучению. Потому что если подвергнуть интегральную схему, в которой записаны команды для робота, действию радиоактивного излучения, то эти команды сами собой меняются. Поэтому почти что все роботы, направленные в зону аварии, не вернулись обратно.
В конце января 2017 года Токийская энергетическая компания ввела в бетонное основание (пьедестал) корпуса высокого давления реактора камеру-зонд. Выяснилось, что в металлическом рабочем помосте прямо под реактором образовалась большая дыра и что расплавившееся ядерное

топливо прошло сквозь дно корпуса высокого давления и опустилось ещё ниже.

Однако в результате этого исследования выяснилась ещё более важная вещь.

Для человека однозначно смертельна доза излучения в 8 Зв. Прямо под реактором уровень излучения составляет 20 Зв в час – это более опасный уровень. Но не доходя до пространства под реактором было зафиксировано излучение в 530-650 Зв[*3]. Соответственно, высокий уровень радиации был зафиксирован не внутри цилиндрического пьедестала, а в пространстве между стенками пьедестала и защитной оболочки.

Токийская энергетическая компания и правительство разработали сценарий, по которому расплавившееся ядерное топливо лежит грудой, как пирожки мандзю, внутри пьедестала и в котором определили, что «через 30-40 лет расплавившееся топливо будет собрано и запечатано в реакторе. Это называется "устранением" аварии».

Но на самом деле расплавившееся ядерное топливо утекает за пределы пьедестала и распространяется снаружи. Правительству и Токийской энергетической компании пришлось переписать «дорожную карту» и заявить о намерении проделать в защитной оболочке отверстие и через него вытаскивать ядерное топливо. Однако не факт, что такая попытка увенчается успехом, так как уровень радиации, которой подвергнутся рабочие, сильно вырастет.

Я с самого начала говорил, что ничего не остается, кроме как закрыть станцию саркофагом, как сделали после аварии на Чернобыльской АЭС в СССР.

Саркофаг на Чернобыльской АЭС прослужил 30 лет и был значительно изношен, так что в ноябре 2016 года был надстроен второй, ещё более крупный по размерам саркофаг. Ожидается, что он прослужит 100 лет. Что делать по прошествии этого времени, неизвестно.

Ныне живущим людям не увидеть устранения последствий аварии на Чернобыльской АЭС.

Тем более не увидеть им устранения последствий аварии и на АЭС Фукусима-1.

Даже если удастся запечатать в реакторе расплавившееся топливо, это не значит, что исчезнет излучение, которое оно испускает. После этого придётся сотни тысяч, а то и миллионы лет поддерживать реактор в безопасном состоянии.

До сих пор ощущается катастрофа, нависшая над окружающей средой в районе АЭС.

В день аварии было объявлено чрезвычайное положение в связи с ядерной аварийной ситуацией[*4]: сначала зона принудительной эвакуации была установлена в радиусе 3 км, затем она была расширена до 10, а затем до 20 километров – люди покинули свои дома только с ручным багажом. Скот и домашние животные были брошены.

Более того, в село Иитате, находившееся в 40-50 километрах от АЭС Фукусима-1, сразу после аварии не поступало никаких объявлений или указаний, и только спустя более чем через месяц по причине чрезвычайного загрязнения была объявлена эвакуация, и вся деревня была выселена.

Что же есть счастье для человека?

Для многих счастье – это возможность только лишь беззаботно, мирно жить с любимым человеком, в окружении семьи, друзей, соседей – жить сегодня, завтра, послезавтра…

Всего этого не стало в одночасье.

Сначала люди, эвакуированные из зоны бедствия, жили в убежищах, например спортзалах, затем ютились по два человека во временном жилье площадью в 4,5 татами, потом их переселили в дома для пострадавших от катастроф или (во временные) дома, предназначенные для сирот. Рассыпались некогда жившие вместе семьи. Прежняя жизнь была разрушена, и даже сейчас люди, не в силах смириться с крахом своих надежд, сводят счёты с жизнью [*5].

И это ещё не всё. Радиоактивное загрязнение вышло за пределы зоны принудительной эвакуации, определённой вследствие чрезмерного заражения этой территории, и, собственно говоря, следовало соответственно расширить зону радиационного контроля.

Зона радиационного контроля – это территория, на которой разрешено находиться только работающему с радиацией и получающему за это деньги персоналу. Однако даже им запрещено пить и есть на этой территории. Разумеется, нельзя и спать. Там нет туалетов, нельзя отправлять

стественные потребности. А правительство под предлогом чрезвычайного положения, в арушение существующих законов оставило в зоне радиоактивного заражения миллионы людей и ринудило их там жить.
рошенные на этой территории люди, среди которых есть и маленькие дети, пьют там воду, едят спят. Над ними, естественно, нависает радиоактивная угроза.
ти люди не уверены в завтрашнем дне. Есть люди, которые целыми семьями, бросив работу, ежали от радиации. Были и случаи, когда, желая защитить от радиации хотя бы детей, тец семейства оставался работать в зоне заражения, а мать эвакуировалась с детьми. Но так азрушается жизненный уклад, разлучаются семьи. Остаться в зоне заражения – значит подорвать доровье, а уехать – всё равно что разбить собственное сердце.
окинутые люди уже более 8 лет живут в каждодневных страданиях.
есмотря на это, в марте 2017 года правительство указало лицам, эвакуированным или амостоятельно покинувшим свои дома, вернуться в загрязненные зоны, уровень радиации в оторых в год не превышает 20 мЗв, и прекратило программу жилищной помощи, до этого ак или иначе поддерживавшую беженцев. Так некоторым пришлось вернуться на заражённые ерритории.

Самое важное сейчас – это восстановление Фукусимы.
3 ситуации, когда ничего не остаётся, кроме как жить там, всем, разумеется, приходится ишь взывать о восстановлении. Эти люди не могут жить в ежедневном страхе. Они хотят забыть о радиоактивном загрязнении, и, к сожалению или к счастью, радиация не видна глазу. Правительство и органы местной власти не жалеют сил, чтобы вынудить людей забыть о беде. Поэтому каждого, кто говорит о неустроенности или загрязнении, они клеймят как препятствующего восстановлению.
Доза радиации в 20 мЗв в год – это предельная доза радиации, изначально одобренная правительством для «работников, связанных с радиацией», к которым когда-то относился и я. Одобрять такую дозу радиации для обычных людей, для которых нет в этом выгоды, – непозволительно. Не говоря уже об уязвимых для радиации детях, которые не в ответе ни за выход японской атомной энергетики из-под контроля, ни за аварию на АЭС Фукусима-1. Недопустимо применять к ним радиационные нормативы для лиц, работающих с источниками излучения.
Однако японское правительство заявляет: «Объявлено чрезвычайное положение в связи с ядерной аварийной ситуацией – это вынужденная мера».
Ещё можно понять, когда режим чрезвычайного положения вводится на день, на неделю, на месяц – в худшем случае на год. Но ведь прошло уже более 8 лет, а режим чрезвычайного положения в связи с ядерной аварийной ситуацией так и не снят. Правительство намеренно делает всё, чтобы люди забыли об аварии на АЭС Фукусима-1, и контролирует СМИ, так что режим чрезвычайного положения в связи с ядерной аварийной ситуацией до сих пор не снят, а многие забыли, что это попирает существующие законы.
Самый опасный радиоактивный загрязнитель окружающей среды – это цезий-137, и период его полураспада составляет 30 лет. Через 100 лет его количество снизится лишь до десятой части от исходного.
Откровенно говоря, в государстве Япония даже через 100 лет будет действовать режим чрезвычайного положения в связи с ядерной аварийной ситуацией.

Во все времена Олимпийские игры служили средством повышения престижа государства.
В последние годы Олимпийские игры стали лакомым куском для корпораций, главным образом строительных, получающих огромную выгоду от крупных строек за государственный счёт в угоду излишне расточительному обществу.
А ведь самое важное сейчас – это приложить все доступные государству усилия к снятию режима чрезвычайного положения в связи с ядерной аварийной ситуацией. Первостепенная задача – спасти людей, продолжающих страдать от последствий аварии. Необходимо защитить от

радиации хотя бы ни в чём не повинных детей.

А правительство всё равно заявляет, что Олимпийские игры – это важно.

Чем дольше сохраняется опасность, тем охотнее власть имущие закрывают на неё глаза. Чтоб заставить общественность забыть о Фукусиме, СМИ будут всё больше подогревать ажиотаж вокру Олимпиады, и придёт время, когда оппонентов Игр назовут «непатриотами».

Так было во время войны.

СМИ передавали только сообщения из Ставки, и почти всё население работало на фронт. Че больше становилось японцев, уверенных в своей национальной исключительности, тем яростне клеймили «непатриотами» и уничтожали противников войны. Но я предпочту быть «непатриотом страны, в которой «Олимпийские игры – это важно», а невинные люди оставлены на произво судьбы.

Авария на АЭС Фукусима-1 останется страшной трагедией на сотни лет.

На многочисленных жертв аварии смотрят искоса, а виновники: Токийская энергетическа компания, правительственные чиновники, учёные, представители СМИ – никто, ни один челове не понёс ответственность. Никто не наказан. Пользуясь этим, они говорят о запуске станции и даже об экспорте АЭС за рубеж.

Олимпиада в Токио пройдёт в стране, в которой введён режим чрезвычайного положения в связи с ядерной аварийной ситуацией.

Страны-участницы и люди, которые приедут на Олимпиаду, разумеется, подвергнутся, с одной стороны, радиационной опасности, а с другой – станут соучастниками преступных действий Японского государства.

[*1] На конференции министров стран-участниц МАГАТЭ по ядерной безопасности, прошедшей через три месяца после аварии, правительство Японии представило доклад «Об аварии на АЭС Фукусима-1, управляемой «Токийской энергетической компанией». Его можно прочитать на главной странице сайта Канцелярии премьер-министра Японии.

Штаб по борьбе с ущербом, нанесённым деятельностью ядерных объектов, «Доклад японского правительства на конференции министров стран-участниц МАГАТЭ по ядерной безопасности – Об аварии на АЭС Фукусима-1, управляемой Токийской энергетической компанией», июнь 2011 года
https://www.kantei.go.jp/jp/topics/2011/iaea_houkokusho.html

[*2] По состоянию на 8 апреля 2019 года объём зараженной воды составил 1136400 тонн. Это в 5 раз больше, чем водоизмещение крупнейшего в мире пассажирского корабля Symphony of the Seas, способного принять на борт более 5400 человек.

*3 Через полгода, в июле 2017 года, Токийская энергетическая компания опубликовала исправленные в сторону понижения данные: прямо под реактором излучение менее 10 Гр в час, в пространстве между стенками пьедестала и защитной оболочки – от 70 до 80 Гр в час.

Международный исследовательский институт проблем вывода из эксплуатации ядерных объектов (IRID), «Токийская энергетическая компания: "О результатах обследования внутренней части второго реактора и замеров уровня радиации"», 27 июля 2017 года
http://irid.or.jp/wp-content/uploads/2017/07/20170728_2.pdf

[*4] Режим чрезвычайного положения в связи с ядерной аварийной ситуацией вводится премьер-министром Японии на основании закона об особых мерах по борьбе с ущербом, нанесенным деятельностью ядерных объектов, в случае возникновения крупной аварии на ядерном объекте. Впервые он был объявлен 11 марта 2011 года в 7 часов 3 минуты из-за аварии на АЭС Фукусима-1 и с тех пор не был снят.

[*5] За 8 лет, прошедших со дня Великого восточно-японского землетрясения, количество эвакуированных в префектуре Фукусима составило 41299 человек (в префектуре Иватэ – 3666 человек, в префектуре Мияги – 2083 человека); число погибших вследствие землетрясения – 2250 человек (в префектуре Иватэ – 467 человек, в префектуре Мияги – 928 человек); число самоубийств, связанных с последствиями землетрясения – 104 человека (в префектуре Иватэ – 50 человек, в префектуре Мияги – 57 человек).

中国語／Chinese]

福岛事故与东京奥运会

——无视真相即是犯罪

小出裕章（原京都大学核反应堆实验所助理教授）

2011年3月11日，东京电力福岛第一核电站遭到大地震及海啸的袭击，核电站全站停电。

专家们一致认为，核电站的全站停电“最有可能导致核电站引发毁灭性事故”。正如他们所预测的那样，在那之后，福岛第一核电站的核反应堆熔毁，将大量放射性物质扩散到了周边地区。
日本政府向国际原子能机构提交的报告书[*1]指出，这起事故向大气中排放的铯137多达1.5×10的16次方贝克勒，相当于向广岛投下了168颗原子弹。向广岛投下的原子弹所产生的辐射足以骇人听闻，而这次，日本政府公布的被排放到大气中的辐射竟是广岛原子弹的168倍之多。
虽然1、2、3号机的核反应堆在这起事故中熔毁，但在堆芯中却残留了合计7×10的17次方贝克勒的铯137，相当于向广岛投下了8000颗原子弹。其中，相当于168颗原子弹所产生的铯137被排放到了大气中，加上流入海洋的铯137，据估测，现在已有相当于1000颗广岛原子弹所产生的铯137被排放到了外界环境中。
铯137是铀裂变形成的核裂变产物之一，也是福岛事故中给人类带来最大威胁的放射性物质。也就是说，堆芯中放射性物质较多的部分，现在依然残留在福岛第一核电站已损坏的核反应堆厂房中。如果堆芯继续熔解，将导致带有铯137的放射性物质再次被排放到外界环境中。为了防止其发生，在事故过去8年多的现在，人们仍在向着存在于某处的已熔毁的堆芯不停注水。因此在核反应堆中，每天仍贮存着数百吨辐射污水。

东京电力已在核电站内建造了将近1000个储水罐以储存污水，其储水总量已超过100万吨[*2]。
然而，核电站的面积有限，储水罐也并非可以无限地增建下去。在不久的将来，东京电力将不得不把辐射污水排向海里。
毋庸置疑，当前的首要任务是尽可能地保证熔毁的堆芯处于安全状态，然而，在事故过去8年多的今天，我们就连熔毁的堆芯的确切位置，以及处于怎样的状态都不得而知。这是因为我们无法前往事故现场。
如果引发事故的是火力发电厂，问题就会简单许多。虽然起初火灾可能会持续数日，但只要控制了火势，就能前往事故现场，甚至可以调查现场情况，对发电厂进行修复，使其重新开始运作。
然而，如果事故是由核电站所引起的，那么人只要进入事故现场，就无法生还。
虽然政府和东京电力试图以机器人代替人进入事故现场，但机器人无法抵挡核辐射。这是因为，如果输入命令的IC芯片暴露于辐射下，命令本身就会被改写。因此，到目前为止被派往事故现场的机器人几乎无一成功返回。
2017年1月末，东京电力在支撑核反应堆压力容器的混凝土底座内部插入了类似胃镜的远程摄像头。经勘测发现，位于压力容器正下方的钢铁作业台上出现了大洞，熔毁的堆芯穿透压力容器的底部，落到了更深的地方。
不仅如此，这次调查还发现了一个更重要的事实。
如果全身遭受了8戈瑞的辐射，人将必死无疑。而压力容器的正下方的辐射可达到每小时20戈瑞。仅是如此就已经十分惊人了，而经过测量却发现，在到达压力容器正下方之前，首先要遭受530或650戈瑞辐射的攻击[*3]。
并且，测量到该高辐射的地点并非圆筒形底座的内部，而是底座壁和安全壳壁之间。
在东京电力和日本政府的构想中，熔毁的堆芯会像小馒头一样堆积在底座内部，他们声称“到30至40年后为止，将回收所有熔毁的堆芯并将其封闭于容器内，以宣告此次事故的结束”。
而实际情况却并非如此。熔解的核燃料泄漏、飞散到了底座外侧，政府和东京电力不得不改写“蓝图”，声称要在安全壳侧部开洞取出堆芯。然而，如果进行这项作业，作业人员将遭受大量的辐射，这显然是不可行的。

我在事发当时就曾提出，要解决这一问题，唯一的办法就是效仿苏联在切尔诺贝利核电站事故时采取的措施，以石棺将其彻底封闭。

切尔诺贝利核电站的石棺在30年之后变得破旧不堪。2016年11月，在其外部又新建造了第二个巨大的石棺，据说可维持100年。但是在那之后还能采取怎样的措施却不得而知。
当下活着的人们当中，无人能见证切尔诺贝利事故的结束。
更何况是福岛事故，就算活着的人们全都死去，也不会迎来终结。
即使假设成功将熔毁的堆芯封闭在容器里，其产生的核辐射也不会消失。而在那之后的几十万年至100万年的时间里，都必须要一直安全地保管容器。
直到现在，核电站周围的环境依然遭受着沉重的打击。
事故发生当天，政府发布了“核能紧急事态宣言”[*4]，下达了强制避难的指示，并将避难范围从最初的3公里依次扩大至10公里、20公里。人们在离开家时仅携带了随身行李，而家畜及宠物则都被遗弃。
此后，距离福岛第一核电站40至50公里、在刚发生事故时没有收到任何警告或指示的饭馆村，在事发一个多月后被告知该地区已受到严重污染。遵照避难指示，全村人都搬离了村庄。

“人们的幸福”究竟是什么？

对于大多数人来说，“幸福”就是与家人、伙伴、邻里、恋人共度的安祥时光，是明天、后天，以及大后天都能一直持续下去的平平淡淡的每一天。
可是，这样的幸福却在某一天突然被剥夺了。
避难的人们起初暂住在体育馆等避难场所，之后又从两人一室、7.4平米的简易住宅，搬到赈灾住宅及无偿公共临时住宅。在这个过程中，曾经一起生活的家人四散分离，至今仍有人因为生活被彻底摧毁而感到绝望，选择亲手结束自己的生命[*5]。
不仅如此，因为遭受严重污染而被迫强制避难的地区的外部，也出现了大范围本应被指定为“辐射管理区域”的受污染地区。
“辐射管理区域”是指除以管理辐射为手段获利者，以及从事辐射相关工作者之外的人皆不准入内的场所。而且，即便是从事辐射相关工作者，在进入辐射管理区域后，也被禁止饮水进食，就寝就更不用说了。辐射管理区域内甚至没有洗手间，人在里面也无法排泄。可是，政府却声称现在处于紧急事态，不顾既往的法律法规，将数百万人遗弃在受污染的地区，强迫他们在那里生活。
包括婴儿在内，被遗弃的人们的吃、喝、睡都在那里进行，他们必然承受着遭受辐射的风险，想必都深感不安。为了避开辐射，有人放弃了工作，举家避难。有的父亲希望至少保证孩子不遭受辐射，选择自己留在受污染地区继续工作，让妻儿去别处避难。但是这样一来，不仅日常生活会遭到破坏，就连家庭也面临着四分五裂。留在受污染地区，会留下身体上的创伤，可如果选择避难，又要承受精神上的打击。
被遗弃的人们在事发后的8年多的时间里，没有一天不与苦恼为伴。
现状如此，而政府却在2017年3月要求曾听从指示避难或主动避难的人们返回全年辐射不超过20毫西弗的受污染地区。并且，政府还中止了此前一直提供的不足以解决问题的住宅补贴，使得一些人不得不被迫回到受污染地区。

在福岛，灾后重建工作是眼下的重中之重。
如果除了在那里生存以外别无选择，大家能做的必然只有祈祷灾后重建。而人们在每天都处于恐慌的状态下，是无法生存的。大家都想要忘记污染仍然存在，而核能——既是幸事也是不幸——是肉眼所看不到的。政府及地方公共团体正努力引导大家忘记事实。因此，如果提及污染或不安，就会遭到非议，被认为是阻碍了灾后重建工作。
我过去也是“从事辐射相关工作者”。全年20毫西弗的辐射量是政府向从事辐射相关工作者制定的允许受到的辐射上限。但是对于不会因为遭受辐射而获得任何利益的人们来说，为他们设定同样的上限这一行为本身是不可原谅的。更何况婴儿和儿童对于辐射很敏感，他们对于日本核能的失控以及福岛事故又分明不需要承担任何责任。无论如何，都不应该以向从事辐射相关工作者制定的标准要求这些无辜的人。
但是日本政府的态度却是“现在要遵照‘核能紧急事态宣言’，发生这种事也是无可奈何的”。

如果紧急事态只持续一整天、一整周、一个月，甚至很遗憾地持续了一整年之久，那么这样的声明也并非不可理解。但实际情况却是，事故发生后已经过去了8年多，“核能紧急事态宣言”时至今日依然没有被解除。

政府有意让人们忘记福岛事故，媒体对此也闭口不谈。“核能紧急事态宣言”现在仍然无法解除，迫使大多数国民忘记了政府无视既往的法律法规的事实。

污染环境的放射性物质的罪魁祸首是铯137，其半衰期为30年。也就是说，即便过去100年，也只不过能减少到现在的十分之一。

我们应认清这样一个现实：在日本这个国家，即使是100年之后，“核能紧急事态宣言”也仍将行使其效力。

不论在哪个时代，奥运会都会作为发扬国威的一种手段被人们利用。

近年来，不必要的公共设施建了又拆、铺张浪费的社会风气，以及以建筑公司为中心、依靠这种手段获得厚利的一批企业都对奥运会趋之若鹜。

但是，现在更为重要的，是举国上下为能够尽快解除“核能紧急事态宣言”而努力。救济因福岛事故而遭受苦难的人们才是当务之急。我们至少要保护无辜的儿童，使他们远离辐射。

然而，这个国家却仍在强调奥运会的重要性。

内部危机越多，有权者就越试图忽视危机的存在。并且，为了让人们忘记福岛事故，媒体在今后会进一步煽动奥运会的热潮。反对奥运会的人被扣上非国民的帽子的日子想必也会到来。

二战时就是如此。

媒体只报道大本营发布的消息，几乎全国都参与协助了战争。越是自认为是“优秀的日本人”的人，就越容易视反对战争的邻居为非国民，将其定罪并彻底排除。如果这个国家一面置无辜之人于不顾，一面又宣扬奥运会的重要性，那我宁愿做一个非国民。

福岛事故承载着巨大的悲剧，今后也将以100年为单位存在下去。

造成这起事故的东京电力、政府官员、学者以及媒体相关人士无一不冷眼旁观着众多受难者，无人为此事负责，更无人受到处罚。他们庆幸着，试图让已停止运作的核电站重新开始工作，甚至还提出要将核能输出到国外。

东京奥运会将在“核能紧急事态宣言”仍然生效的情况下举行。

前来参加奥运会的各个国家的人们不仅要冒着遭受辐射的风险，同时，他们也将成为这个犯下罪行的国家的帮凶。

[*1] 事故发生3个月后，在就核能安全问题召开的IAEA（国际原子能机构）理事会会议上，日本政府提交了一份题为《关于东京电力福岛核电站的事故》的报告书。该报告书的全文可于首相官邸的网站中浏览。

核能灾害对策总部《日本政府向就核能安全问题召开的IAEA理事会会议提交的报告书》2011年6月

https://www.kantei.go.jp/jp/topics/2011/iaea_houkokusho.html

[*2] 截至2019年4月8日，污水总量已达到「113.64万吨。相当于5艘“海洋交响号”—

—可载客5400人的世界最大的邮轮的重量。

[*3] 东京电力在半年后，也就是2017年7月将其测量的每小时的核辐射值修改为更小的数字：容器正下方的辐射为每小时10戈瑞以下，而其外部的底座壁和安全壳壁之间则为70或80戈瑞。

国际废反应堆研究开发机构 东京电力《关于2号机安全壳内部的调查——辐射剂量率的确认结果》2017年7月27日

http://irid.or.jp/wp-content/uploads/2017/07/20170728_2.pdf

[*4]“核能紧急事态宣言”在核能设施发生异常重大的事故时，基于《核能灾害对策特别措置法》，由内阁总理大臣发布。该宣言因东京电力福岛第一核电站的事故，于2011年3月11日晚上7点3分首次被发布，且现在仍然生效。

[*5] 东日本大地震发生后的8年时间里，福岛县的避难者人数达到4万1299人（岩手县3666人、宫城县2083人），间接死于灾害者2250人（岩手县467人、宫城县928人），因间接受灾而自杀者104人（岩手县50人、宫城县57人）。

[アラビア語／Arabic]

حادثة فوكوشيما وأولمبياد طوكيو

تجاهل الحقائق جريمة

كويده هيرواكي (أستاذ في مخبر تجارب المفاعل النووي في جامعة كيوتو سابقا)

في الحادي عشر من أيار عام 2011 ضرب زلزال قوي رافقه تسونامي المفاعل النووي الأول لمحطة توليد الطاقة الكهربائية في فوكوشيما. ونتج عنه توقف الكهرباء لكامل المنطقة.

يتفق رأي الاختصاصين أن السبب الأول ذا الاحتمالية الكبيرة الذي يقف وراء انقطاع الكهرباء في كامل المنطقة هو حدوث حادث كارثي أصاب محطة التوليد النووية. ووفقا لهذا التصور، فقد انصهر المفاعل في محطة توليد الكهرباء النووية الأول في فوكوشيما، وقد نتج عنه انتشار كمية كبيرة من المواد المشعة في البيئة المحيطة.

وفقا للتقرير [1] الذي قدمته الحكومة اليابانية للوكالة الدولية للطاقة الذرية فإنه خلال الحادث تم انبعاث 1.5 مضروب ب 10 مرفوعة للقوة 16من نظير السيزيوم 137 انتشرت في الهواء وهذه الكمية تعادل 168 ضعف ما نتج من القنبلة الذرية التي أصابت هيروشيما. إن الكمية التي انتشرت من المواد المشعة من قنبلة هيروشيما مخيفة للغاية، فكيف الأمر بانتشار168 ضعف من المواد المشعة في الهواء وفقا لقول الحكومة اليابانية!

في ذلك الحادث تم انصهار كل من المفاعل الأول والثاني والثالث. وإجمالي نظير السيزيوم 137 الموجود داخل المفاعلات يصل 7 مضروب ب10 مرفوعة للقوة 17 أي ما يقارب 8000 ضعف القنبلة الذرية في هيروشيما. ومن هذه الكيمة تم انتشار 168 ضعف، ومع إضافة ما تم انتشاره في البحر فقد تصل الكمية إلى 1000 ضعف القنبلة الذرية في هيروشيما حتى الوقت الراهن.

يعتبر نظير السيزيوم 137 أحد العناصر المشعة الناتجة من عملية الانشطار النووي لعنصر اليورانيوم أكبر خطر على الإنسان. أي أن الجزء الأكبر من العناصر المشعة في قلب المفاعل مازالت في حجرة المفاعل النووي المتحطم التابع لمحطة توليد الكهرباء النووية الأولى في فوكوشيما. وفي حال انصهار قلب المفاعل فإن العناصر المشعة بما فيها نظير السيزيوم 137 ستنتشر مرة أخرى في البيئة. وللوقاية من ذلك، فإنه منذ الحادث وحتى الآن بعد مضي 8 سنوات يتم صب الماء على قلب المفاعل المنصهر بشكل مستمر، ومن أجل ذلك ستتجمع مئات الأطنان يوميا من المياه الملوثة شعاعيا.

قامت شركة طوكيو للطاقة الكهربائية داخل المناطق التابعة لها بصنع ما يقارب ألف خزان لتجميع المياه الملوثة شعاعيا، وقد تجاوزت الكمية الكلية للمياه الملوثة ما 100 طون[2].

ومن المعلوم أن المنطقة التابعة للشركة محدودة وكذلك إضافة خزانات جديدة له حدود، وبالتالي فإنه في المستقبل القريب ستضطر الشركة إلى صب المياه الملوثة في البحار.

من المؤكد أن أهم شيء هو وضع قلب المفاعل المنصهر في حالة آمنة ولو نسبيا. ولكن بعد مضي ثماني سنوات وحتى الآن لا نعرف ما هو وضع قلب المفاعل المنصهر ولا مكانه، وذلك لأنه لا يمكن الذهاب إلى المنطقة.

وفي حالة أن الحادث نتج عن محطة توليد الكهرباء الحرارية فالموضوع بسيط، إذ قد تستمر الكارثة لبضعة أيام ثم تهدأ ويصبح من الممكن زيارة المكان، حيث سيتم فحص حالة الحادث وإصلاحه ثم إعادة المحطة للعمل، ولكن عندما يكون الحادث قد وقع في محطة توليد كهرباء باستخدام الطاقة النووية فذلك يعني أن أي شخص يذهب إلى ذلك المكان ستكون النتيجة الموت حتماً.

فكرت الحكومة وشركة طوكيو لتوليد الطاقة الكهربائية في إرسال الرجال الآلية بدلا عن الإنسان، ولكن المشكلة أن الرجال الآلية ضعيفة المقاومة تجاه الأشعة، وذلك لأن الشرائح الإلكترونية للدارات المتكاملة التي تمت برمجتها إذا تعرضت للأشعة فإن الأوامر البرمجية ستتغير. ولهذا فإن معظم الرجال الآلية التي تم إرسالها إلى منطقة الحادث لم تعد سالمة.

في نهاية الشهر الأول من عام 2017 قامت شركة طوكيو لتوليد الكهرباء بإدخال آلة تصوير وتنظير دقيقة يتم التحكم بها عن بعد إلى داخل القاعدة الإسمنتية التي تحمل حوجلة المفاعل، وقد تبين أنه توجد فتحة في السقالة الفولاذية تحت حوجلة المفاعل وأن الحوجلة مثقوبة وينزل منها قلب المفاعل المنصهر إلى الأسفل.

ولكن فقد تبين من خلال هذه العملية ما هو أهم من ذلك.

إن الإنسان إذا تعرض إلى 8 زيفرت من المواد المشعة بكامل جسده فسيكون موته محتما. وقد بلغت كمية الإشعاعات تحت حوجلة المفاعل مباشرة إلى 20 زيفرت خلال الساعة الواحدة. وعلى الرغم أن هذه كمية هائلة من الإشعاعات، فإن كمية الإشعاعات قبل أن تصل إلى تلك النقطة تقدر بما يقارب 530 زيفرت أو 650 زيفرت [3]، بالإضافة إلى أن هذه الكمية تم تقديرها ليس داخل القاعدة الأسطوانية، وإنما في المنطقة الفاصلة بين جدار القاعدة الأسطوانية وجدار الوعاء المحيط بها.

وقد قامت شركة طوكيو لتوليد الكهرباء بكتابة سيناريو يفيد بأنه سيتم تجميع قلب المفاعل المنصهر داخل القاعدة الحاملة لحوجلة المفاعل على شكل مادة لزجة.أي أن (بعد من 30 إلى 40 عاما، سيتم تجميع القلب المنصهر ضمن وعاء ومن ثم يتم إغلاقه. وهكذا تكون نهاية الحادث)

لكن الحقيقة هي أن الوقود النووي قد تسرب إلى خارج القاعدة الحاملة منتشرا. وعليه قامت الحكومة وشركة طوكيو لتوليد الكهرباء بتغير خريطة الطريق حيث بدأت تقول إنه سيتم ثقب جدار الوعاء المحيط من الخاصرة ويتم سحب المادة من داخلها. ولكن للقيام بمثل هذا العمل

سيؤدي إلى تعريض العاملين إلى كمية ضخمة من المواد المشعة وبالتالي فإنه لا يمكن القيام بذلك.

منذ البداية قلت أنه لا يوجد حل إلا كما قام الاتحاد السوفيتي بعمله في حادثة تشيرنوبيل وذلك بحجز المكان بصندوق حجري.
ولكن بعد مضي 30 عاما على حادثة تشيرنوبيل فإن الصندوق الحجري قد تهدم لذا قاموا بتغليفه بصندوق حجري آخر في عام 2016 في شهر نوفمبر (تشرين الثاني). ويبلغ عمر الصندوق الثاني 100 عاما، وبعدها لا أعرف ما هي الطريقة التي يمكن عملها.
اليوم لا يمكن لأي شخص من الناس أن يرى نهاية حادثة تشيرنوبيل. أضف إلى ذلك أنه بعد موت كل الناس الأحياء حاليا فإن مشكلة حادثة فوكوشيما لن تنتهي، فحتى ولو تم احتواء المادة الناتجة عن انصهار قلب المفاعل في وعاء بشكل مؤقت، فإن ذلك لا يعني أبدا أنه تمت إزالة المواد المشعة، إذ يتوجب الحفاظ على هذا الوعاء في مكان آمن من مئات السنين إلى مليون سنة.
و حتى الآن تستمر المأساة الهائلة في البيئة المحيطة بمحطة توليد الكهرباء.
ففي اليوم الأول من الحادث تم إعلان حالة الطوارئ للطاقة الذرية[4]، وتوسعت المنطقة التي تم أمر الإخلاء القسري منها من 3 كيلو متر في البداية إلى 10 كيلو متر ثم 20 كيلو، حيث نزح الناس وحملوا فقط ما يستطيعون حمله بأيديهم مغادرين منازلهم. وتركت وراءها أي مواشي أو حيوانات كانت لديها.
ثم بعد ذلك تم الابتعاد من 40 إلى 50 كيلو. وبالنسبة لقرية إيتاته التي لم تتلق أي تحذير بعد الحادث، فقد تم بعد مضي أكثر من شهر إخلاؤها بشكل كامل بسبب التلوث الهائل فيها.

بماذا ياترى يمكن لنا أن نعرف السعادة بالنسبة للإنسان.

بالنسبة إلى كثير من الناس فإن السعادة هي استمرار الحياة اليوم وغدا وبعد غد وبعده بدون مكروه لكل من الأهل والأصدقاء والجيران والأحبة.
في ذلك اليوم وبشكل مفاجئ تغير ذلك كله.
لجأ الناس في البداية إلى الملاجئ كالصالات الرياضية ثم إلى أبنية مؤقتة يبلغ ارتفاعها 7 متر مربع لكل شخصين. ثم إلى مبانٍ تم بناؤها بعد الكارثة ثم انتقلوا إلى بيوت تم تمويلها من قبل الدولة بعد الكارثة. وخلال تلك الفترة تشتت أفراد الأسرة الذين تعودوا العيش معا. حيث تدمرت الحياة بشكل كامل فهناك من انتحر وسط فقدان الأمل ومازال هناك من يفكر بالانتحار أيضا[5].
ليس ذلك فحسب، فإنه ستتولد أيضاً منطقة كبيرة من المنطقة المحيطة بالمنطقة التي تم إخلاؤها بشكل قسري بسبب التلوث الشعاعي، وهي من حيث المبدأ لا بد من اعتبارها (منطقة خاصة للأشعة خاضعة للإدارة).
يقصد ب(منطقة خاصة للأشعة خاضعة للإدارة) هي المنطقة التي يسمح فقط للعاملين فيها بالدخول إليها وهم البالغون الذين يتعاملون مع المواد المشعة والذين يتقاضون رواتب لقاء عملهم. وفي تلك المنطقة فإنه حتى العمال لا يسمح لهم بالطعام أو الشراب، وبالتأكيد فإن النوم ممنوع ولا توجد حمامات ولا يسمح بالتبول، ولكن الحكومة الآن تحت ظل إعلان الحالة الطارئة وبشكل مخالف للقواعد المعمول فيها حتى الآن، فإنها قد تركت ملايين الناس يعيشون في تلك المنطقة.
من بين هؤلاء الناس رضع يشربون الماء ويأكلون الطعام وينامون، ومن البديهي أنهم معرضون لخطر التعرض للأشعة النووية. كل هؤلاء يشعرون بالحيرة، فمنهم من هرب من التعرض للأشعة النووية مع كل أهله وترك عمله، ومنهم من أرسل أطفاله لحمايتهم من التعرض للأشعة وبقي العاملون في المنطقة الملوثة بهدف العمل وكسب المعيشة، ومنهم من أرسل الأطفال والأمهات. ولكن بذلك تكون الأسر قد تدمرت وكذلك الحياة، فإن بقي في المنطقة الملوثة فسيتعرض للخطر في جسده وإن رحل فسيكون الألم والخطر في قلبه.
هؤلاء الناس منذ أكثر من 8 سنوات كل يوم يعيشون وهم يحملون الألم في صدورهم.
ومع ذلك كله، قامت الحكومة في أيار من عام 2017 بتوقيف الدعم المالي للسكن الذي قدمته للاجئين الذين تم إخلاؤهم أو هؤلاء الذين نزحوا من تلقاء أنفسهم، وذلك في المناطق التي لا تزيد فيها نسبة التلوث عن 20 ملي زيفرت مطالبة إياهم بالعودة إلى تلك المناطق.

إن أهم شيء اعتبر مهما الآن في فوكوشيما هو عودة الحياة إليها.
فإن كان الوضع فيها أنه لا يوجد خيار سوى العيش فيها، فبالتأكيد ستكون أمنية الجميع أن تعود الحياة إليها، ولكن لا يمكن للإنسان العيش هناك بدون أن يحمل في صدره مشاعر الخوف، وستكون هناك الرغبة بنسيان التلوث. ولا نعرف أنه لحسن الحظ أم لسوئه أن المواد المشعة لا يمكن رؤيتها بالعين المجردة. ولهذا تعمل الحكومة ومجالس الإدارة المحلية بشكل متعمد على اعتماد أسلوب النسيان، حيث تنتقد كل من يتكلم عن التلوث والقلق على أنه شخص عائق لعودة الحياة الطبيعية.
إن التعرض لكمية 20 ملي زيفرت في السنة يعتبر أول مرة تسمح فيه الحكومة اليابانية بالتعرض لهذه النسبة للعاملين في مجال الطاقة الذرية مثلي، ولكن أن يتعرض أناس عاديون ليست لهم أية مصلحة أو عمل لهذه الأشعة فهذا شيء يصعب غفرانه. أضف إلى ذلك أن الرضع أكثر عرضة للخطر . هؤلاء ليست لهم أية مسؤولية في حادثة فوكوشيما أو في تقدم مجال توليد الطاقة باستخدام الطاقة النووية، فلا يمكن لهؤلاء أن يتعرضوا إلى نفس الكمية التي تعتبر مرجعا للعاملين في مجال الطاقة الذرية.
ولكن الحكومة اليابانية تقول (كون الحالة طارئة فلا يوجد حل بديل).
الحالة الطارئة يمكن أن تكون يوما أو أسبوعا، إن كانت شهرا فسيعتبر ذلك أمر سيء، وإن امتدت لعام فيمكن بطريقة ما تفهم الأمر. ولكن

المشكلة أنه في الواقع قد مضى 8 سنوات ومازال الوضع تحت الحالة الطارئة للطاقة الذرية.
تتعمد الحكومة أن تنسي الناس حادثة فوكوشيما وتطبق بإحكام على الإعلام، فحتى الآن ما زال الوضع حالة طارئة والذي يعد مخالفا للقواعد المعتبرة وكثير من المواطنين قد نسوا ذلك.
إن المادة الأكثر تلويثا للبيئة هي نظير السيزيوم 137 والتي يبلغ نصف عمرها 30 عاما، فحتى ولو مضى 100 عام فسيبقى 10% منهما.
في الحقيقة فإن هذا البلد اليابان حتى ولو بعد مضي 100 عام ستبقى تحت الحالة الطارئة للطاقة الذرية.

تم استخدام الألعاب الأولمبية في كل عصر بهدف رفع سمعة الدولة.
وفي هذه الأيام تهدف الشركات من خلال الأولمبياد إلى إقامة مبانٍ تكلف المجتمع تكاليف باهظة تتركز عوائدها بأيدي شركات البناء وحسب.
ولكن المهم حاليا هو وبشكل فوري وعاجل إلغاء حالة الطوارئ المعلن عنها وعلى الحكومة أن تبذل ما بوسعها لذلك. لا بد من مساعدة المتضررين من حادثة فوكوشيما والتي تستمر مأساتهم، وعلى أقل تقدير يجب حماية الأطفال الذين لا ذنب لهم من التعرض للأشعة.
ولكن مع ذلك ، فإن هذه الحكومة تعتبر أن الأولمبياد مهمة.
في الداخل بقدر ما يحمل الوضع من مخاطر بقدر ما يصرف أصحاب السلطة أنظارهم عن تلك المخاطر. ثم من أجل نسيان حادثة فوكوشيما يقوم الإعلام بالتركيز على الأولمبياد ومن يعارض الأولمبياد فربما سيأتي يوم يتهم فيه بأنه غير وطني.
كما حدث ذلك أثناء الحرب. وبقدر ما رأى الياباني نفسه أنه وطني، فإنه اتهم من يعارض الحرب على أنه غير وطني وقام بقتله أيضا. ولكن إن كانت الدولة ستهتم بالأولمبياد وتهمل الناس الذين لا ذنب لهم فسأكون سعيداً أن أُعتبر غير وطني.

إن حادثة فوكوشيما بما تحمله من مأساة عظيمة ستستمر كما هي 100 عام.
وهؤلاء الذين كان لهم دور في تلك الحادثة من شركة طوكيو لتوليد الكهرباء ومن لهم علاقة من الحكومة وكذلك العلماء وكل من له علاقة من الإعلام لم يتحمل واحد منهم المسؤولية وهم ينظرون إلى الضحايا بلا اكتراث، ولم تتم محاسبتهم. وبالنسبة لهم فإن الأمر الجيد هو إعادة تشغيل محطة توليد الطاقة الكهربائية باستخدام الطاقة الذرية، بل وأيضا تصديرها إلى الدول الأخرى حسب قولهم.

افتتاح أولمبياد طوكيو في دولة ما زالت تحت الحالة الطارئة للطاقة النووية.
ولهذا فإن من يشارك من دول هم بالتأكيد معرضون لخطر التعرض للأشعة من طرف ومن طرف آخر فهم يساهمون في دعم من كان له دور في جريمة الدولة.

[1] بعد ثلاثة أشه من الحادث، قامت الحكومة اليابانية بتقديم تقرير (حادث محطة توليد الطاقة النووي في فوكوشيما التابع لشركة طوكيو لتوليد الطاقة الكهربائية) خلال الاجتماع الوزاري الذي تم عقده بخصوص أمان الطاقة النووية IAEA (منظمة الطاقة النووية الدولي). يمكن الإطلاع على كامل نص التقرير من خلال موقع ريئس الوزراء على الرابط التالي:
المركز الرئيسي لإجراءات السلامة من كوارث الطاقة النووية (التقرير المقدم من الحكومة اليابانية خلال الاجتماع الوزاري لمناقشة أمان الطاقة النووية التابع لمنظمة الطاقة النووية الطولي IAEA في الشهر السادس من عام 2011
https://www.kantei.go.jp/jp/topics/2011/iaea_houkokusho.html

[2] وصلت كمية المياه الملوثة شعاعيا إلى 1136400 طون في الثامن من نيسان عام 2019. إذا تم مقارنة هذه الكمية مع أضخم سفينة في العالم (الملقبة ب لحن البحار) والتي تتسع إلى 5400 راكب فإن كمية المياه الملوثة تبلغ خمسة أضعافها.

[3] بعد مضي نصف عام أي في الشهر السابع من عام 2017 قامت شركة طوكيو للكهرباء بتعديل التقييمات إلى 10 زيفرت مباشرة تحت الحوجلة، وفي المسافة بين الجدار الخارجي للحوجلة والجدار المغلف لها من 70 إلى 80 زيفرت وذلك لكمية الإشعاعات خلال ساعة واحدة.
المعهد الدولي لأبحاث اتلاف المواد الشعاعية شركة طوكيو للكهرباء (فحص الجدار الداخلي لحوجلة المفاعل الثاني ~ نيتجة نسب الكميات) في 27 من الشهر السابع عام 2017.
http://irid.or.jp/wp-content/uploads/2017/07/20170728_2.pdf

[4] قام ريئس الوزراء بإعلان حالة الطوارئ للطاقة النووية بسبب وقوع حادث ضخم في منشأة للطاقة النووية وبالاعتماد على قانون الحالات الخاصة لإجراءات الكوارث المتعلقة بالطاقة النووية. وتم إعلانه رسميا في 11 من شهر آذار عام 2011 في الساعة 7 وثلاث دقائق ليلا وذلك نتيجة للحادث الذي وقع في محطة توليد الطاقة النووية الأولى في فوكوشيما التابعة لشركة طوكيو للكهرباء. والقانون نافذ حتى الآن.

[5] منذ وقوع الحادث في شرق اليابان وإلى 8 سنوات بلغ عدد اللاجئين إلى 41299 شخص في فوكوشيما (و3666 شخص في إيواتيه و 2083 في مياغي) وعدد الضحايا إلى 2250 في فوكوشيما (و467 شخص في إيواتيه و 928 في مياغي) وعدد المنتحرين بسبب الكارثة في فوكوشيما إلى 104 (و50 شخص في إيواتيه و 57 في مياغي)

あとがき

Epilogue

Nachwort

Épilogue

Epílogo

Послесловие

后记

ملحق

人間は神ではなく、過ちから無縁でない。また、事故や故障から無縁な機械はない。原子力発電所(原発)は、広島原爆に比べて数千倍の放射能を抱える機械である。人間が望もうと望まなかろうと、原発が破局的な事故を起こす可能性はある。それを知った時、私は一刻も早く原発を廃絶したいと考えた。そして私の人生のほぼ半世紀をそのために生きてきた。しかし、私の願いは叶えられず、フクシマ事故は起きた。

筆舌に尽くしがたい被害と被害者が生まれた。一方、原発の破局的事故は決して起こらないと嘘をついてきた国や東京電力は、誰一人として責任を取ろうとしないし、処罰もされていない。絶大な権力を持つ彼らは、教育とマスコミを使ってフクシマ事故を忘れさせる作戦に出た。そして、東京オリンピックのお祭り騒ぎに国民の目を集めることで、フクシマ事故をなきものにし、一度は止まった原発を再稼働させようとしている。フクシマ事故が起きた当日に発令された「原子力緊急事態宣言」は事故から8年経った今も解除できないままである。しかし、国民のほとんどはその事実すら知らない。

そんな時、イタリア在住の楠本淳子さんが私に一文を書くよう勧めてくれた。彼女はそれを世界各国のオリンピック委員会に送るという。私自身は原子力の旗は決して振らなかった。しかし、原子力の場にいた人間として、フクシマ事故に重い責任が私にはあると思う。そこで私は「フクシマ事故と東京オリンピック」という文章を書いた。その文章に今回、径書房が目を止めてくれ、7カ国語に翻訳したうえで、出版してくれることになった。私は自分の本を出すということに興味がなく、本を出すために文章を書いたことはない。しかし、止むに止まれぬ思いで書いた文章を、多くの人の手に届けて下さるというお申し出はありがたいことと思う。私にこの文章を書くように勧めてくれた楠本淳子さん、そして、編集者の藤代勇人さん、径書房の原田純さんのお力添えに感謝する。

2019年4月26日（チェエルノブイリ事故33周年の日に）

小出 裕章
Hiroaki Koide

[英語／English]

Human beings are no God, they belong to a species doomed to make mistakes. Likewise, no machine can rule out eventual defects and incidents.

Nuclear plants are a type of machinery which contains thousands of times more radiation than the Hiroshima nuclear bomb. Regardless of our best intentions, it cannot be denied that nuclear plants are strictly intertwined with the risk of potentially catastrophic failures.When I finally admitted this to myself, I started to wish that nuclear plants be dismantled as quickly as possible. To this I dedicated most of my professional life, amounting to almost half a century now. Unfortunately, my wish did not come true; a nuclear accident did happen at the Fukushima nuclear plant, an accident that caused indescribable damage and left an ominous number of casualties behind.

TEPCO and the Japanese government have lied for years by concealing the risk associated with nuclear plants and ruling out any risk of failure; and even now neither of them has admitted its responsibilities or else has held accountable for the accident that has occurred.On the contrary, taking advantage of their immense powers, they set up a real propaganda campaigns by exploiting the mass media and manipulating the educational system, in an attempt to make people forget about the Fukushima accident.Lately, they have been drawing people's attention to the 2020 Tokyo Olympics, a genuine example of pageantry, diverting their attention from the Fukushima accident, as if it had never happened, while they attempt to reactivate other nuclear plants that had previously been shut off. Through all of this, the state of nuclear emergency, declared by the government the very day of the Fukushima accident has not been lifted yet, after eight long years; and the majority of people are not even aware of it.

Miss Junko Kusumoto, who lives in Italy, has prompted me to write an article which she then sent to all the branches of the International Olympic Committees (IOC) in the world. Although I never participated in promoting the development of nuclear plants, as a professional who has worked in the field of nuclear science I feel that I do have my share of responsibility in relation to the accident that occurred at the Fukushima nuclear plant. For this reason, I agreed to her request and I wrote an article entitled: " The disaster in Fukushima and the 2020 Tokyo Olympics ". The Komichi Publishing Company stumbled upon my article and decided to publish it in seven languages. Since I have never had any interest in publishing books, I have also never written anything to this purpose; but I cannot thank them enough for their gesture, since the very deeply-felt article that I wrote can now be delivered to the hands of many more people than I could possibly dream to reach out to. I now take this opportunity to thank Miss Junko Kusumoto for prompting me to write this article, as well as the editor, Mr.Hayato Fujishiro,and Miss Jun Harada, editor of the KOMICHI-SHOBO, for their kind supporto.

April 26th 2019, on the 33rd anniversary of the Chernobyl Disaster

[ドイツ語／German]

Menschen sind keine Götter und daher vor Fehlern nicht gefeit. Unfall- und störungsfreie Maschinen gibt es ebenso wenig. Atomkraftwerke (AKW) sind Maschinen mit einer im Vergleich zur Hiroshima-Atombombe tausendfachen Radioaktivität. Ob der Mensch will oder nicht, AKW können katastrophale Unfälle verursachen. Als ich das verstanden hatte, wollte ich, dass sie so schnell wie möglich abgebaut werden. Dem habe ich fast mein halbes Leben gewidmet. Doch ging mein Wunsch nicht in Erfüllung, es kam zur Katastrophe in Fukushima.

Die Schäden und Opfer, die daraus hervorgingen, sind kaum in Worte zu fassen. Von denen aber, die uns seitens des Staates und von TEPCO belogen haben, ein katastrophaler Unfall werde in einem AKW nicht geschehen, hat bisher niemand Verantwortung übernommen, niemand wurde bestraft. Ausgestattet mit grenzenloser Macht, gingen sie daran, den Unfall in Fukushima mit Hilfe der Bildungseinrichtungen und der Massenmedien vergessen zu machen. Mehr noch, die Aufmerksamkeit des Volkes auf das Festtrubel der Olympischen Spiele lenkend, soll diese Katastrophe in Vergessenheit geraten und die einmal herunt-

ergefahrenen AKW wieder in Betrieb genommen werden. Die am Tag der Dreifachkatastrophe verkündete „Erklärung des atomaren Ausnahmezustands" ist auch heute, nach acht Jahren noch in Kraft, doch die meisten Japaner wissen nicht einmal das.
Unter diesen Umständen bat mich die in Italien lebende Kusumoto Junko, darüber etwas zu schreiben. Das wolle sie dann an alle Nationalen Olympischen Komitees schicken. Ich habe die Fahne der Atomenergie zwar niemals hochgehalten. Doch war ich ja in einer solchen Einrichtung tätig, weshalb ich eine schwere Schuld für die Katastrophe in Fukushima empfinde. Und so habe ich diesen Text Die Katstrophe von Fukushima und die Olympischen Spiele in Tōkyō 2020 geschrieben. Darauf ist dann der Verlag Komichi Shobō aufmerksam geworden, der ihn in sieben weitere Sprachen übersetzen und veröffentlichen wollte. Ich selbst bin an einem eigenen Buch nicht interessiert und habe bislang nie einen Text um seiner Publikation willen geschrieben. Dennoch bin ich überaus dankbar dafür, dass dieser aus vollster Überzeugung verfasste Text nun in die Hände so vieler Menschen gelangt. Mein Dank gilt Frau Kusumoto Junko, die mich dazu ermunterte diesen Text zu schreiben, aber auch dem Herausgeber, Herrn Hayato Fujishirō und Herrn Harada Jun vom Verlag Komichi Shobō.
26. April 2019 (33. Jahrestag von Tschernobyl)

[フランス語／French]

L'être humain n'est pas Dieu et n'est pas à l'abri des erreurs. De plus, il n'y a pas de machine qui est exempte des accidents et des pannes. Une centrale nucléaire est une machine qui émet plusieurs milliers de fois plus de radiations que la bombe atomique d'Hiroshima. Que les humains le veuillent ou non, il est possible que la centrale nucléaire provoque un accident catastrophique. Quand j'en ai compris, j'ai pensé que nous devions éliminer la centrale nucléaire dès que possible. Et j'ai consacré presque un demi-siècle de ma vie à cela. Malgré cela, mon souhait n'est pas exaucé et l'accident de Fukushima a eu lieu.
Il est impossible de décrire la gravité des dommages et des victimes. En revanche, personne du gouvernement ou de TEPCO qui avait menti, affirmant que la catastrophe nucléaire ne se produira jamais, n'assume aucune responsabilité et n'a pas été puni. Ils détiennent le pouvoir énorme et ont servi l'éducation et les médias pour faire oublier cet accident de Fukushima. Et, en attirant l'attention du public lors du festival des Jeux Olympiques de Tokyo, ils tentent d'effacer de mémoire sur l'accident de Fukushima et de réactiver les centrales nucléaires qui étaient fermées. La "Déclaration d'urgence nucléaire" publiée le jour même de l'accident de Fukushima n'est pas encore levée, huit ans après l'accident. Cependant, la plupart des gens ne le savent même pas.
C'est alors qu'Atsuko Enomoto, qui vit en Italie, m'a conseillé d'écrire à ce sujet. Elle m'a dit qu'elle l'enverrait aux Comités des Jeux Olympiques de tous les pays membres. Je n'ai jamais pris la position de soutenir l'énergie nucléaire, mais en tant que personne travaillant dans le secteur de l'énergie nucléaire, je dois une lourde responsabilité sur l'accident de Fukushima. J'ai donc écrit le texte intitulé "L'accident de Fukushima et les Jeux Olympiques de Tokyo". Et cela a attiré l'attention du Komichi Shobo qui a voulu le traduire en sept langues et publier. Je n'ai jamais pensé à publier mon propre livre, je n'ai donc jamais écrit les textes qui pourraient être pour. Cependant, j'apprécie énormément l'offre de transmettre mon message, sur lequel j'ai mis tous mes pensées et convictions, à la portée de beaucoup de gens. Je suis reconnaissant à Junko Kusumoto de m'avoir encouragée d'écrire ce texte, à Hayato Fujishiro, éditeur, ainsi qu'à Jun Harada de Komichi Shobo.
26 avril 2019 (à l'occasion du 33e anniversaire de l'accident de Tchernobyl)

[スペイン語／Spanish]

Las personas no son dioses y no están ajenas a los errores. Ni siquiera las máquinas están ajenas a los accidentes y las fallas de su funcionamiento. Una central de energía atómica (nuclear) es una máquina que tiene miles de veces más radiación que la bomba atómica de Hiroshima. Lo quieran o no las personas, siempre existe la posibilidad de que ocurra un accidente catastrófico. Cuando yo comprendí esto, pensé deinmediato que las plantas nucleares debían cerrarse cuanto antes. Y para lograrlo he dedicado casi la mitad de siglo de mi vida. Sin embargo, sin que mi deseo se cumpliera, ocurrió el accidente de Fukushima.

No existen palabras para describir el daño y las víctimas resultantes. El gobierno y TEPCO, que tantas mentiras dijeron sobre lo imposible que sería que ocurra un accidente así en una central nuclear, no fueron ni responsabilizados ni castigados. Ellos, que tanto poder tienen, impulsaron una estrategia para hacernos olvidar del accidente de Fukushima mediante el uso del sistema educativo y de los medios de comunicación. Y así, al hacer que todas las miradas de la población se concentren en las festividades de las Olimpíadas, el accidente de Fukushima se verá enterrado y las centrales nucleares detenidas volverán a ser puestas en funcionamiento. Mientras, la Declaración de Emergencia Nuclear pronunciada el mismo día del accidente sigue en vigencia, hoy, ocho años después del siniestro. Mucha gente todavía desconoce este hecho.

Kusumoto Junko, que entonces vivía en Italia, me sugirió que escribiera algunas palabras al respecto. Dijo que ella enviaría el texto a los comités olímpicos de todo el mundo. Yo jamás levanté una bandera a favor de la energía nuclear. Pero como persona que trabajó utilizando esta última, creo que también tengo una pesada responsabilidad de lo sucedido en Fukushima. Y fue por esa razón que escribí aquel texto titulado “El accidente de Fukushima y las Olimpíadas de Tokio”. Ahora, gracias al interés de la editorial Komichi Shobō, dicho texto será publicado en siete idiomas distintos. Yo no tengo un particular interés en que un libro mío salga a la venta, ni escribí aquel texto para que se transformara posteriormente en un libro. Sin embargo, no puedo dejar de recordar y de agradecer a todas las personas que me ayudaron a escribirlo. Estoy profundamente agradecido a Kusumoto Junko por su sugerencia, y al editor Fujishiro Hayato y a Harada Jun, editor de Komichi Shobō.

26 de abril de 2019 (en el aniversario número 33 del accidente de Chernóbil)

[ロシア語／Russian]

Люди не боги и не застрахованы от ошибок. Нет и механизма, в котором не может случиться поломки или аварии. Атомная электростанция – это механизм, обладающий радиационным потенциалом, в тысячи раз превосходящим потенциал ядерного взрыва в Хиросиме. На что бы ни надеялись люди, всегда есть вероятность разрушительной аварии на АЭС. Когда я понял это, я подумал, что хочу, чтобы мы как можно скорее отказались от ядерной энергии. Я положил на это почти половину собственной жизни. Но мои мольбы не были услышаны – произошла авария на АЭС Фукусима-1.

Масштаб разрушений и страдания жертв аварии не поддаются описанию. Но правительство и Токийская энергетическая компания, лгавшие нам о том, что катастрофы на АЭС не может произойти, – никто не понёс ответственности, никто не наказан. Облечённые огромной властью, они через СМИ и систему образования делают всё, чтобы заставить людей забыть о трагедии. Более того, отвлекая людей праздничной шумихой вокруг Олимпийских игр в Токио, они пытаются похоронить память о произошедшем на Фукусиме и снова запустить электростанцию. Объявленный в день аварии режим чрезвычайного положения в связи с ядерной аварийной ситуацией действует уже 8 лет и до сих пор не отменён. И почти никто не знает об этом.

В это время живущая в Италии Кусумото Дзюнко посоветовала мне написать об аварии статью. Она сказала, что разошлёт её в Национальные олимпийские комитеты стран мира. Я сам ни в коем случае не выступал за ядерную энергию. Но как человек, который бывал на атомных электростанциях, я считаю и себя тоже ответственным за аварию на АЭС Фукусима-1. Так я написал статью «Авария на АЭС Фукусима-1 и Олимпийские игры в Токио». В этот раз на неё обратило внимание издательство Комити Сёбо, которое решило издать её на 7 языках. Лично я не заинтересован в издании книги и написал статью не ради этого. Однако я считаю, что это очень хорошо, что многие люди смогут прочитать статью, которую я написал, будучи во власти собственных мыслей. Я благодарю за поддержку Кусумото Дзюнко, посоветовавшую мне написать эту статью, редактора Фудзисиро Хаято и Харада Дзюн из издательства Комити Сёбо. 26 апреля 2019 года (день 33-ей годовщины аварии на Чернобыльской АЭС)

【中国語／Chinese】

人非圣贤，孰能无过。机器亦然。没有哪台机器可以保证永远不引发事故或出现故障。核电站就是这样一台“机器”，它所拥有的核能相当于在广岛投下原子弹时产生的核能的数千倍。无论人们愿意与否，核电站引发毁灭性事故的可能性都是存在的。得知这一点后，我便产生了一定要尽快废除核电站的想法。并且，在几乎半个世纪的人生当中，我一直都在为此而努力。然而，我的愿望还未实现，福岛事故就发生了。

事故带来的灾难及受灾人数之多无法用语言形容。另一方面，一直谎称核电站不会带来毁灭性事故的政府和东京电力之中，没有一个人站出来承担责任，也没有人受到处罚。他们大权在握，利用教育及媒体的手段，企图使人们忘记福岛事故的存在。此外，他们还引导国民将目光转向东京奥运会这一盛典，以此掩盖福岛事故存在的事实，并试图让一度停止运作的核电站重新开始工作。福岛事故发生当天发布的“核能紧急事态宣言”在事故过去8年后的今天仍然没有被解除。然而，绝大多数的国民就连这一事实也不知晓。

当时，居住在意大利的楠本淳子女士建议我提笔撰文，再由她发送给世界各国的奥委会。我个人绝不支持推广核能，但是，作为曾经身处核能相关环境中的一员，我认为福岛事故的沉重责任也有我的一份。于是，我写下了《福岛事故与东京奥运会》一文。径书房注意到了这篇文章，将它翻译成7国语言并出版了。我个人对于出版自己的书一事并没有兴趣，也从来没有为了出书而写文章。但是我很感激径书房提议将我以久久不能平静的心情写下的文章，送到许许多多的人手中。在此，我要对建议我撰写这篇文章的楠本淳子女士、给予我协助的编辑藤代勇人先生，以及径书房的原田纯女士致以衷心的感谢。

2019年4月26日（于切尔诺贝利事故33周年之日）

【アラビア語／Arabic】

إن الإنسان ليس إلهاً وليس معصوماً عن الخطأ. وليست هناك آلة بدون عطل أو حادث. إن محطة توليد الطاقة الكهربائية باعتماد الطاقة النووية هي آلة تحمل بداخلها آلاف الأضعاف من المواد المشعة بالمقارنة مع قنبلة هيروشيما النووية. وبرغبة الإنسان أو بدونها فإن احتمالية حدوث حادث مدمر لتلك المحطة موجود. ومنذ أن عرفت ذلك وأنا أفكر في إنهاء تلك المحطة بشكل فوري وعاجل. وقد أمضيت نصف عمري من أجل ذلك. ومع ذلك لم يتحقق حلمي ووقع حادث فوكوشيما.

وقد نتج عن هذا الحادث ضحايا وأضرار لا يمكن وصفها، وفي المقابل فإن الحكومة وكذلك شركة طوكيو لتوليد الكهرباء كذبتا بقولهما أنه لا يمكن وقوع حادث مدمر للمحطة، فلم يتحمل المسؤولية أي منهما، ولم تتم محاسبة أحد. وهؤلاء الذين لديهم السلطة الكاملة يقومون باستخدام الإعلام والتعليم كاستراتيجية لجعل الناس ينسون حادثة فوكوشيما، ويعملون على إعادة تشغيل المحطة التي توقفت، ويجمعون أنظار المواطنين إلى أولمبياد طوكيو وكأن حادثة فوكوشيما لم تحدث.

ومنذ اليوم الأول لوقوع الحادث تم الإعلان عن حالة طوارئ للطاقة النووية ولكن بعد مضي أكثر من 8 أعوام مازالت الحالة حالة طارئة إلا أن معظم الناس لا يعلمون ذلك.

في ذلك الحين أشارت علي كوسوموتو جونكو المقيمة في إيطاليا بكتابة مقال، حيث قامت هي بإرساله إلى لجنة الأولمبياد الدولية. وعلى الرغم من أني لم أرفع راية الطاقة الذرية يوما، فإني أجد نفسي حاملا مسؤولية كبيرة لكوني عملت في مجال الطاقة الذرية. ولذلك كتبت مقالاً بعنوان (حادث فوكوشيما وأولمبياد طوكيو). حيث لفت المقال انتباه دار نشر كوميتشي شوبو التي قامت بترجمته إلى سبع لغات ونشره في كتاب.

ليس لدي اهتمام بنشر كتاب ولم أكتب المقال بهدف تأليف كتاب، ولكن لم أستطع منع نفسي من كتابة المقال الذي أتمنى أن يصل إلى أيدي كثير من الناس. وأتوجه بالشكر إلى كل من كوسوموتو جونكو التي أشارت علي بكتابته وكذلك المحرر فوجيشيرو هاياتو وهارادا جون من دار كوميتشي شوبو.

في 26 من نيسان عام 2019 (الذكرى 33 عاما على مرور حادث تشيرنوبيل)

写真説明&クレジット／Photo caption and credit

002-003 2013年9月7日にIOCでプレゼンテーションを行う安倍晋三首相。ユーチューブに流れる首相官邸作成の動画より。
Prime Minister Shinzo Abe giving a presentation at IOC on September 7, 2013 IOC. From a video clip on You Tube created by the Prime Minister's Office.

006-007 福島第1原子力発電所が立地する双葉町に1988年に設置された原発PR看板には、当時小学6年生だった大沼勇治さんが学校の宿題で考案した「原子力明るい未来のエネルギー」という標語が掲げられていた。写真は大沼さん自身の目。看板は2015年12月に撤去され、文字のパネルは福島県立博物館に保管されている。2015年4月12日。写真=中筋純
The slogan, "Nuclear power is the energy for a bright future" written as a school homework by Yuji Onuma, then a 6th grade elementary school student, was put on the signboard of Fukushima Daiichi Nuclear Power Plant in 1988 in Futabamachi where the nuclear power plant is located. The eye in the photo is of Mr. Onuma himself. The signboard was removed in December 2015, and the panel with the slogan is kept in storage at Fukushima Museum. Apr. 12, 2015. Photo by Jun Nakasuji.

014-015 地球／Globe [CG] MIyabi-K/PIXTA

016-018 日本列島／Japanese archipelago [CG] Dream Finder/PIXTA

022-023 福島第1原発近くに位置する双葉北小学校の教室に貼られたままのカレンダー。あの日から時は止まったまま。2015年4月13日。写真=中筋純
A calendar left on a classroom wall at Futaba-kita Elementary School near the nuclear power plant. Time has stopped since that day. April 13, 2015. Photo by Jun Nakasuji.

024-025 事故前の福島第1原発全景。写真=東京電力ホールディングス
A full view of Fukushima Daiichi Nuclear Power Plant before the accident. Photo provided by Tokyo Electric Power Company Holdings.

026-027 福島第1原発を襲う津波。2011年3月11日15時37分頃。写真=東京電力ホールディングス
Tsunami reaching Fukushima Daiichi Nuclear Power Plant. Around 15:37, Mar. 11, 2011. Photo provided by Tokyo Electric Power Company Holdings.

030-031 浸水による全電源喪失で炉心の冷却が不可となった結果、津波襲来から丸1日後の3月12日15時36分頃に原子炉1号機が、14日11時01分頃には3号機が水素爆発。写真はネット上で拡散されている爆発時の映像。
As a result of the station blackout from the flood caused by the tsunami, a hydrogen explosion took place at reactor 1 at 15:36, Mar. 12, one day after the arrival of tsunami, and at reactor 3 at 11:01 on Mar. 14. Photo taken from a video capturing the scene of the explosion and was spread on the Internet.

032-033 原子炉3号機の爆発から3分後に撮られた衛星写真。中央の白煙を上げているのが3号機。写真=Digital Globe/Abaca/アフロ
Satellite photo taken 3 min. after the explosion at reactor 3. The middle building spewing white smoke is reactor 3. Photo provided by Digital Globe/Abaca/Aflo

036-039 米軍によって撮影された広島原爆で発生したきのこ雲。1945年8月6日。写真=広島平和記念資料館
Mushroom cloud from the atomic bomb explosion in Hiroshima taken by the U.S. Army. Aug. 6, 1945. Photo provided by Hiroshima Peace Memorial Museum.

046-047 無人飛行機によって撮影された原子炉3号機（左）と4号機。2011年3月24日。写真=エア・フォート・サービス/東京電力ホールディングス
Reactors 3 (left) and 4 taken from an unmanned aircraft, Mar. 24, 2011. Photo provided by Air Photo Service, Tokyo Electric Power Company Holdings.

048-049 福島第一原発全景。刻々と増える汚染水タンクの置き場所は限界に達しつつある。2018年12月。写真=秋田放射能測定室べぐれでねが
Full view of Fukushima Daiichi Nuclear Power Plant. Land space for placing the growing number of water tanks with contaminated water is about to reach the limit. Dec. 2018. Photo provided by Akita Radiation Measuring Station (Beguredenega)

058-059 1986年4月に発生したチェルノブイリ原発事故による放射能汚染を封じ込めるために造られた4号炉の石棺。事故直後の高線量下での突貫工事だったため各所に崩落個所が目立つ。2013年2月13日。写真=中筋純
Temporary shelter structure built around reactor 4 of the Chernobyl Nuclear Power Plant for containing radiation after the accident in Apr. 1986. Photo shows many places that are falling apart due to hasty construction in a highly radioactive environment just after the accident. Feb. 13, 2013. Photo by Jun Nakasuji.

060-061 新しい石棺（右・ドーム状）を建設中のチェルノブイリ原発。当初の予定より工期は大幅に延び、2019年夏にようやく完成。耐用年数は100年といわれている。2014年2月15日。写真=中筋純
Chernobyl Nuclear Power Plant where the new safe confinement (right/dome shaped structure) is being built. Initial construction period was extended considerably and the confinement was finally completed in summer of 2019. It is said to last for 100 years. Feb. 15, 2014. Photo by Jun Nakasuji.Chernobyl Nuclear Power Plant where the new safe confinement (right/dome shaped structure) is being built. Initial construction period was extended considerably and the confinement was finally completed in summer of 2019. It is said to last for 100 years. Feb. 15, 2014. Photo by Jun Nakasuji.

066-067 低く垂れこめた朝もやのなかに浮かび上がる何列もの高圧電線。向かう先は関東地方。やがてもやが晴れると福島第1原発が姿を現す。福島県大熊町の丘より。2016年11月20日。写真=中筋純
Numerous lines of high-voltage cables stretching toward Kanto region appear in the morning haze. As the haze clear out, the image of Fukushima Daiichi will also appear. Nov. 20, 2016. Taken from a hill in Okumamachi, Fukushima. Photo by Jun Nakasuji.

070-071 福島第1原発から3 キロほどにある大熊町のJR大野駅前通り。かつては数多くの商店が軒を連ねていた。帰還困難地区にありながら、現在は特定復興再生拠点区域として2020年3月のJR常磐線再開に合わせて除染作業が急ピッチで進む。しかしそれは、事故後の時間の経過で痛みが激しい店舗家屋にとっては解体を意味することでもある。2018年8月20日。写真=中筋純
JR Ono Station street in Okumamachi, 3km. from Fukushima Daiichi Nuclear Power Plant. There used to be many shops lined up on this street. Though it is located within the difficult-to-return zone, it is also designated as specified reconstruction and revitalization base where decontamination work is fast in progress to meet the reopening of JR Joban Line. However, such designation also means tearing down shops and houses that have become deteriorated over time since the accident. Aug. 20, 2018. Photo by Jun Nakasuji.

072 [上] 南相馬市小高区の金房小学校校庭の遊具。150年近い歴史をもつ同校は震災後警戒区域内となり、現在は小高小学校にて4校合同での教育活動を行っている。2013年10月10日。写真=中筋純
[top] Playground equipment on the grounds of Kanabusa Elementary School in Odaka, Minamisoma City. With a history of nearly 150 years, the school was designated as evacuation zone after the earthquake disaster, therefore it is now operating together with 3 other schools at Odaka Elementary School. Oct. 10, 2013. photo by Jun Nakasuji.

072 [下] 富岡町の小学校の下駄箱に残された運動靴。あのとき高学年だった子どもたちは今や大学生や社会人になっている。2019年7月5日。写真=中筋純

oottom] Students' sneakers left inside the shoe shelf at an lementary school in Tomiokamachi. Students who were in higher rades back then are now university students or at work. July 5, 019. Photo by Jun Nakasuji.

73 [上] 富岡町にある小学校の教室の黒板には、担任教諭が一時帰還した際に残したと思われる、「またいつか教室で会いましょう!!」というメッセージが書かれていた。この学校校舎は近い将来解体され歴史に幕が閉じられる。2014年4月13日。写真=中筋純
top] A message on the blackboard of an elementary school in omiokamachi believed to have been left by the classroom teacher when making a temporary return reads, "Let's meet again someday n this classroom!" The school is expected to be taken down in the ear future, marking the end of its history. Apr. 13, 2014. Photo by un Nakasuji.

)73 [下] 震災と原発事故が発生した時期は小学校の卒業式シーズンに重なったため、多くの学校の講堂は式の準備の状態のまま時が止まっている。2019年1月21日。写真=中筋純
[bottom] Since the earthquake and nuclear accident overlapped with the season for elementary school graduation, many schools had made preparation for the ceremony at their auditoriums and the time has stopped there. Jan. 21, 2019. Photo by Jun Nakasuji.

078-079 原発事故で帰還困難となった地域には3500頭ほどの飼い牛が取り残され、その多くは餓死したり殺処分されたりしたが、生き残った牛は野生化した。そのため保健所は放射能汚染の拡散防止のため囲い込み柵を設置して捕獲した。浪江町の「希望の牧場・ふくしま」はそんな捕獲対象の牛を引き受け、出荷できないにもかかわらず飼育し続けている。2013年11月11日。写真=中筋純
Some 3,500 cattle were left in the area designated as difficult-to-return zone due to the nuclear accident, and although many died from starvation or killing, those that survived became wild animals. Because of this, health centers set up a fence to capture the animals to prevent spread of radioactive contamination. "Kibo bokujo Fukushima," a ranch in Namiemachi, has accepted the captured cattle and continue raising them, though they cannot be sent to the market. Nov. 11, 2013. Photo by Jun Nakasuji.

080-081 富岡町のスーパーマーケット。新学期間近ではなやいでいたはずの売り場には制服姿のマネキンが横たわっていた。2015年10月11日。写真=中筋純
Supermarket in Tomiokamachi. A mannequin dressed in school uniform is left lying down in the store filled with merchandise for the coming new school season. Oct. 11, 2015. Photo by Jun Nakasuji.

088-089 福島市のシンボル、信夫山の中腹に造られた仮置き場に積み上げられた「除去土壌」という名の放射性廃棄物。福島原発から50キロ以上離れた福島市は避難指示区域外とされたものの、市内全域で面的除染が行われ、各所に除去土壌が保管されることとなった。2018年8月20日。写真=中筋純
Bags of radioactive waste referred to as "removed soil" are piled up at a temporary storage site on the hill of Mt. Shinobu, Fukushima City's symbol. Although Fukushima City, over 50km. from Fukushima Daiichi Nuclear Power Plant, was not included in the evacuation zone, decontamination works were conducted, leaving the removed radioactive soil in storage at various places in the city. Aug. 20, 2018. Photo by Jun Nakasuji.

090-091 富岡町の町営球場の時間的変遷。2014年より本格化した除染作業で出た大量の廃棄物はフレコンバッグに詰められグラウンドに山積みされた(090[上]:2014年11月9日)。その後正式な仮置き場が確保されると移送が始まり(091[上]:2015年5月13日)、2016年5月にはその姿はきれいに消えた(090[下]:2016年5月9日, 091[下]:2017年4月18日)。写真=中筋純
How Tomiokamachi's public baseball stadium changed over time. Full-scale decontamination work began from 2014 and the removed soil were packed into flexible container bags and piled up on the stadium ground. (90 [top]: Nov. 9, 2014) Eventually the bags began to be removed after a temporary storage site was officially secured. (91[top]: May 13, 2015) And in May 2016, the bags were all gone (90[bottom]: May 9, 2016, 91[bottom]: Apr. 18, 2017) Photo by Jun Nakasuji.

092-093 福島県立美術館前の芝生広場。地中から顔を出すキノコのようなものは、ガス抜きのための塩ビパイプ(写真中央)。その存在が地中に大量の廃棄物が埋められていることを物語る。2018年10月20日。写真=中筋純
Lawn field in front of the Fukushima Prefectural Museum of Art. Vinyl chloride pipes (center of photo) that stick out from the ground like mushrooms are for releasing gas from underground. This shows there are huge amount of radioactive waste still buried underground. Oct. 20, 2018. Photo by Jun Nakasuji.

102-103 第125回IOC総会で2020年五輪の東京開催が決定した瞬間、歓喜に湧く安倍首相(右から3人目)ほか東京五輪招致委員会の面々。2013年9月7日。写真=代表撮影/AP/アフロ
Prime Minister Abe (third from right) and other members of the Tokyo 2020 Bid Committee cheer with joy the moment Tokyo was announced as the 2020 Olympics site at the 125nd IOC meeting. Sept. 7, 2013. Photo provided by AP/Aflo.

104-105 東京都中央区晴海に建設中のオリンピック選手村。大会終了後は5000万円台~1億円超の価格で分譲マンションとして売り出され、1万2000人が暮らす街に生まれ変わるのだという。2019年7月3日。写真=毎日新聞社
Athlete's village under construction in Harumi, Chuo-ku, Tokyo. It is said the rooms will be offered as condominiums for 50~100 million yen after the Olympics, and the village will become a community with a population of 12,000. July 3, 2019. Photo provided by Mainichi Shimbun.

106-107 東日本大震災から8年。プレハブの仮設住宅の通路で話し込む住民たち。福島県南相馬市。2019年3月10日。写真=毎日新聞社
Eight years from the Great East Japan Earthquake. Residents of pre-fabricated temporary housing chat in the corridor. Minamisoma, Fukushima Prefecture. Mar. 10, 2019. Photo provided by Mainichi Shimbun.

108-109 福島県浪江町の住宅地に迫る勢いの廃棄物仮置き場。2015年4月12日。写真=中筋純
Temporary storage site in Namiemachi, Fukushima that is continuing to expand and about to reach a residential area. Apr. 12, 2015. Photo by Jun Nakasuji.

114-115 富岡町仏浜海岸地域には一時期町内中のありとあらゆる除染廃棄物が集積されたが、新たな仮置き場や原発至近に中間貯蔵施設ができると順次搬送され、やがてなにもなくなった。しかしそれは消滅したのではなく、置き場所が変わっただけに過ぎない。2015年10月12日。写真=中筋純
Hotokehama beach area in Tomiokamachi was once a place where all kinds of contaminated waste were brought in. However, when a temporary storage site was built as well as a mid-term storage facility near the nuclear power plant, the waste was removed until eventually the area was empty. This, however, does not mean the waste disappeared, but merely that the storage location changed. Oct. 12, 2015. Photo by Jun Nakasuji.

150-151 006-007の写真は撮影者・中筋純氏主催の「流転 福島展」のキーアイコンとして使用され、その後の巡回展では来場者のメッセージボードとなった。写真展を訪れた全国各地の人々の福島に対する様々な想いが刻まれている。写真=中筋純
Photo on p.6-7 was used as the main icon of photographer Jun Nakasuji's exhibition titled Ruten Fukushima (Fukushima in a state of flux) and became a visitors' message board in the following exhibitions. Various opinions and thoughts toward Fukushima are expressed by people from around Japan who saw the photo exhibition.

日本は原発を悪いと思っていても、他国へ売っているなんて。自国でも稼働させている

地球と人の安心安全を(まりかけて)、核を取るなんてばかげてる!

世界中の原子炉を廃炉に!

脱原発!

あまい言葉にだまされないで!!

オリンピックを原発を東京に!! 原発のTOKYOを作ろう!! 安全なのでしょう?

WERE YOU THERE...?

明るい未来のTOKYOを

原発反対派も賛成派も、笑顔で生きていける社会を作りたい。

やすらかに。いのちを紡いで

自然がいい!!

事実を目にする。人事ではない。

福島を返して!

少欲知足

脱原発

ふるさとを返して!

今自分に必要な分だけあれば、充分幸せ。欲を求めるときりがない。やがて自身をほろぼす。

= 再生可能エネルギー

見えないものを見えるように

一人一人が抱える問題。

いとなみを返して!!

NO NUKE

思い出してみる 2017.2.19

人類、最大最悪の犯罪

完全なんて作れないことを知っている

Damai di dunia

原発事故

がんばれ!

この作品たちはそんな思いにさせてくれました。

原発不要!

原発反対!! ♡ Tp.

原発はもう要りません!!

いつかまた帰りたい

お正月もお盆も久しぶりの家族との時間

次世代の子供のためにも

災害はおこってしまったからしかたないが

帰りたいな

日本中どこも一緒だよ。ふくしまのこと他人事ではない

原発はいりません。

原発反対

福島の現状を聞いて

しっかり見て

全ての原子力に反対。

子供に責任ある未来をつないで行きたい

人間は疾病だ!

この世界を認識

清水克彦

この原発事故に生き甲斐をすべて奪われた

許せない

原発はろくでなし!!

反核

福1のセキュリティーはイスラエルがやったの?

ろくでなし発電だ!! 坂本修一

緑の野菜。きれいな水

再稼働させない。

絶対に。

原子力明るい

No Nukes

緑のわりには

原発が廃止することのみに意味がある。

原発いらない!

人間絶滅

広島、長崎、チェルノブイリそして福島

もう、これ以上、原発はいりません!!

原発で利益を得る人達

植物って命が強い!

平和になりますように

全ての生きている命に優しく幸せになりますように

I love ふくしま!!

独を見習いましょう!

世界には原発はいらない

原発反対!!

No Nukes!

人は欲深いが人間の理性をしっかり信じたい!

Where are you going?

ひろしまがんばれ

今生きている若者や、これから生まれてくる人達も

原発はんたい!!

震災前に想っていた願い…!

人間はアホや!!

No more Fukushima!

平和一番

何をみるのか? 何をみないのか? 私たちの選択は

原発はいらない!

平和な、安全な未来のために。

この悲しい出来事を経験に!!

来てよかったです Chii

対価が大きすぎる。誰がこの犠牲を取るのだろうか? ふるさとを失うことをもっと掘り下げて考えるべきだ。

地球は信じます

「野となれ山となれ」

原発不要

政権を支えているのは? わたしたち、みんな。

2016.3.14. 圭

明るい未来を

これからの日本は廃炉の

福島

がんばってる
あなた！
私！

〈あなたは私
私はあなた〉

人は核を操れない！いい世界を

5年経ってもまだまだ

風化させない!!

「原発は安全」神話の

あの日を決して忘れない

充分がんばってきているから
がんばらなくって出来るコツコツコツコツ
無理なくやっていこう

始めましょうネ…

想いがいつか
力へ変わりますように

この素晴らしい地球を
活断層の上にある日本の土地
即川内原発は止めてくれ――

原発の消滅を…
世界の平和を願います。

二度とくり返さないで!!

いのります

眼に見えないものが
1番大切で、1番怖い

no more 原発！

目撃せよ

小さな力ですが
集まれば、きっと何かができる
そう信じています！

決して忘れない
FUKUSHIMA！

子どもたちに楽しい未来を！

DON'T FORGET NAGASAKI, DON'T FORGET HIROSHIMA! LET US LEARN FROM HISTORY AND LIVE FOR PEACE!

Red Eye

核兵器の

one for all
all for one

原発は絶対いらない

子どもたちに明るい未来を！

日本が原発なくても
電気は出来る!!

原発はいらない

一人一人の力は小さくても できること
皆んなで考えよう！

忘れない！

真の明るい未来がありますように

アメトーーーク!!

祈りをこめて

緑の地球を守りたい!!

すべての核をこの世から
なくしていこう!!

Peace

Love

人間を大切に

原発いりません。

誰か住民に
土下座する奴はいないのか!!

国会に
処理場を作って！
2016.3.13

二度と繰り返さないで

福島を返せっ！

自然を返せっ！

No Nukes！

見なきゃ見えない

しっかり目に焼きつけ
忘れない

再稼働 絶対
許さないっ！

平和の音♪
決して忘れない！

地球はどうなるのか。
今どうすべきか、一人一人考えてほしい

瞳に写ったものは

誰も消せない!!

狭い国・地震の多い国に
原発つくってどうするんだ!!
福島の教訓をいかさないのか？
再稼働許さない!!

原発なくしたい！

この地震大国に、原発もリニアもムリム

二度とくりかえすな!!

愛

ながさきがんばっ

二度とやすれない

原発を容認するなら
まず東京湾に1基作れ！

神はみない

未来にがんばって。
この悲しみを忘れない！

世界皆んなの
幸せを祈って！

現実を知り、それでも

小出 裕章（こいで・ひろあき）
元京都大学原子炉実験所助教。工学修士。
第２次世界大戦が終わった4年後の1949（昭和24）年8月、東京の下町・台東区上野で生まれる。中学生のとき地質学に興味をもち、高校３年までの6年間、ひたすら山や野原で岩石採集に没頭する。68年、未来のエネルギーを担うと信じた原子力の平和利用を夢見て東北大学工学部原子核工学科に入学。しかし原子力について専門的に学べば学ぶほど、原子力発電に潜む破滅的危険性こそが人間にとっての脅威であることに気づき、70年に考え方を180度転換。それから40年以上にわたり、原発をなくすための研究と運動を続ける。2015年3月に京都大学を定年退職。現在は長野県松本市に暮らす。著書に『隠される原子力・核の真実―原子力の専門家が原発に反対するわけ』（2011年11月／創史社）、『原発のウソ』（2012年12月／扶桑社新書）、『100年後の人々へ』（2014年2月／集英社新書）ほか多数。

Hiroaki Koide
Hiroaki Koide is a former assistant professor at Kyoto University Research Reactor Institute (KURRI). Master of engineering. He was born in Ueno,Taito-ku,Tokyo in August 1949, 4 years after the end of World War II. His interest grew in geology during his junior high school years, and transformed into collecting rocks in mountains and fields for 6 years until his high school graduation. Koide's aspirations for the peaceful use of nuclear power directed him to study nuclear engineering within the department of engineering at Tohoku University in 1968. Consequently, the more he studied nuclear power, the more he came to realize that the destructive danger of nuclear power is the very threat to humans. In 1970, he changed his position completely and dedicated himself to research that focused on the elimination of nuclear power plants and has been an activist against nuclear power plants for 40 years. After many years of conducting research, he retired from Kyoto University in March of 2015 and is currently living in Matsumoto-city, Nagano. His other publications include *The Hidden Truth of Nuclear Power and Nuclear Power Plant – Why The Expert of Nuclear Power Is Against Nuclear Power Plant* (November 2011, Sousisha), *The Lies of Nuclear Power Plant* (December 2012,Fusoshashinsho), and *Dear People 100 Years Later* (February 2014,Shueishashinsho).

【7ヵ国語対応】
フクシマ事故と東京オリンピック
The disaster in Fukushima and the 2020 Tokyo Olympics

2019年12月2日　第１刷発行

著者　小出裕章

企画・編集　藤代勇人（紙ヒコーキ舎）

ブックデザイン　櫻井 浩（⑥Design）

翻訳協力　[英語] 楠本淳子, Marco Stocchi Grava (P20-113,144)／西山 彩 (P9,117,著者略歴)／深見明子 (P148-149,帯)
[ドイツ語] Steffi Richter, Felix Jawinski　[フランス語] 原口 健／薛 善子
[スペイン語] Matias Chiappe Ippolito　[ロシア語] Mozebakh Vladimir／鈴木玲子
[中国語] 趙 子男／日本大来株式会社　[アラビア語] Omar Sakka

編集協力　諌山三武／渡辺悦司／桂木 忍／小田明美／原田 純／須藤 惟

スペシャルサンクス　中筋 純 (写真提供 P6-7,58-93,108-151)
[略歴]1966年和歌山県生まれ。東京外国語大学卒業。雑誌編集者を経て写真家として独立。2007年よりチェルノブイリの取材を開始。3.11後は福島の被災地を定期的に訪れその後の「現在」を記録し続けている。2016年より全国30カ所を巡回しながら「流転 福島&チェルノブイリ」写真展を開催。また福島事故に対する想いを表現するアーティストたちによる「もやい展」を主催、震災後10年目にあたる2021年3月には横浜赤レンガ倉庫での開催が決まっている。
◎もやい展〈2021年プロジェクト〉https://www.facebook.com/2019moyai/

発行所　株式会社 径書房（こみち）
〒160-0012　東京都新宿区南元町11-3
電話　03-6629-8518
FAX　0120-531-096
http://site.komichi.co.jp/

印刷・製本　中央精版印刷株式会社

 ISBN978-4-7705-0228-5 C0036